The Physics of Liquid Water

The Physics of Liquid Water

Makoto Yasutomi

JENNY STANFORD
PUBLISHING

Published by

Jenny Stanford Publishing Pte. Ltd.
Level 34, Centennial Tower
3 Temasek Avenue
Singapore 039190

Email: editorial@jennystanford.com
Web: www.jennystanford.com

British Library Cataloguing-in-Publication Data
A catalogue record for this book is available from the British Library.

The Physics of Liquid Water

ISBN 978-981-4877-25-1 (Hardcover)
ISBN 978-1-003-05616-4 (eBook)

Contents

Preface

Water is an important and indispensable material for all living things on earth. As is well known, it is closely related to natural phenomena and life phenomena. Therefore, researchers from various fields have studied the properties of water for centuries and elucidated various interesting properties. As a result, water has become known to have a variety of strange behaviors not found in ordinary materials. Among them, one of its remarkable property is that it exhibits negative thermal expansion below 4°C. A wonderful phenomenon of this property can be seen in areas where during cold winters the surfaces of ponds, lakes, and rivers freeze but the water underneath does not. This allows aquatic organisms to continue their life even in extremely cold winters.

A normal substance expands when heated and contracts when cooled (positive thermal expansion), regardless of temperature T. However, in the case of water, positive thermal expansion occurs as long as $T > 4^\circ$C, and below that, it contracts when heated, and expands when cooled (negative thermal expansion). The thermodynamic mechanism that causes this negative thermal expansion has been a mystery for hundreds of years. Various theories have been advocated so far to solve this mystery, but none could elucidate the mechanism. However, recently, I succeeded in fundamentally elucidating the thermodynamic mechanism.

Here, I would like to emphasize on what we should elucidate to "fundamentally solve the mystery of a thermodynamic phenomenon." Thermodynamic properties of a material are determined by the interactions between the particles that make up the material, and the associated physical quantities can be derived using thermodynamic and statistical mechanics techniques. Therefore, if the relationship between the functional form of interparticle interactions and the thermodynamic phenomenon can be understood using thermodynamic quantities, the mystery of the phenomenon can be fundamentally clarified. In this book, the thermodynamic properties of water have been analyzed from this standpoint and the physical mechanism behind the negative thermal expansion of water has been elucidated. This has been done by explaining the method carefully. Although difficult and complex mathematical formulas have to be used for this purpose but the meaning of the mathematical formulas, the derivation methods, and the computation methods have been explained in as much detail as possible to enhance understanding of the written mathematical formulas.

The search for the thermodynamic properties of materials is actually based on experimental facts. However, there are many things that cannot be obtained by experiments alone, and many facts have been elucidated from numerical experiments using the Monte Carlo method or the molecular dynamics method. At present, most of the experimental facts related to water are advanced enough to be reproduced by numerical experiments using models.

Since a water model is developed mainly to reproduce experimental facts with high accuracy, it incorporates various properties of actual water molecules. However, it cannot be denied that the elucidation of the thermodynamic mechanism that is responsible for the phenomenon is unclear. This is because, in many cases, it is not all of the various properties of water that generate the thermodynamic phenomenon of interest, but only a limited number of these properties. Therefore, in order to understand the physics behind the phenomenon, it is necessary to build a model that incorporates only the properties essential to the phenomenon and remove other miscellaneous properties. A model built in such a way is called a simplified model or core-softened model, which is also discussed in the book.

I would like to thank Jenny Stanford Publishing for giving me the opportunity to publish this book and senior editor Shivani Sharma for the appropriate advices.

Makoto Yasutomi
Autumn 2020

Chapter 1

Statistical Mechanics and Thermodynamics of Fluids

1.1 Statistical Mechanics

Consider a homogeneous isotropic disordered system consisting of $\mathcal{N}$ identical particles. The position vector of each particle is $\mathbf{r}_1, \mathbf{r}_2, \cdots, \mathbf{r}_{\mathcal{N}}$, and the potential energy of the system is $U_{\mathcal{N}}(\mathbf{r}_1, \mathbf{r}_2, \cdots, \mathbf{r}_{\mathcal{N}})$, then the grand canonical partition function Ξ is given by the following formula:

$$\Xi = \sum_{\mathcal{N}=0}^{\infty} z^{\mathcal{N}} Q_{\mathcal{N}}, \tag{1.1}$$

$$Q_{\mathcal{N}} = \frac{1}{\mathcal{N}!} \int d\mathbf{r}_1 \int d\mathbf{r}_2 \cdots \int d\mathbf{r}_{\mathcal{N}} \exp[-\beta U_{\mathcal{N}}(\mathbf{r}_1, \mathbf{r}_2, \cdots, \mathbf{r}_{\mathcal{N}})],$$

$$z = \left(\frac{2\pi m}{\beta h^2}\right)^{3/2} \exp(\beta\mu), \qquad \beta = \frac{1}{kT},$$

where m is the particle mass, h is the Planck constant, k is the Boltzmann constant, T is the absolute temperature, and μ is the chemical potential. The canonical partition function for canonical

The Physics of Liquid Water
Makoto Yasutomi

ISBN 978-981-4877-25-1 (Hardcover), 978-1-003-05616-4 (eBook)
www.jennystanford.com

ensemble $Z_{\mathcal{N}}$ is given by

$$Z_{\mathcal{N}} = \left(\frac{2\pi m}{\beta h^2}\right)^{(3/2)\mathcal{N}} Q_{\mathcal{N}}. \tag{1.2}$$

The following equation holds between the pressure p, volume V, and free energy F:

$$pV = kT \log \Xi, \quad F = -kT \log Z_{\mathcal{N}}. \tag{1.3}$$

Any other thermodynamic quantity can be obtained from Ξ or $Z_{\mathcal{N}}$.

The two-pair distribution function $\rho^{(2)}(\mathbf{r}_1, \mathbf{r}_2)$ for a uniform isotropic fluid (generic term for gas and liquid) is defined by the following equation:

$$\rho^{(2)}(\mathbf{r}_1, \mathbf{r}_2) = \rho^2 g(r_{12}) = \frac{1}{\Xi} \sum_{\mathcal{N} \geq 2} z^{\mathcal{N}} Q_{\mathcal{N}} \rho_{\mathcal{N}}^{(2)}(\mathbf{r}_{12}), \tag{1.4}$$

$$\rho_{\mathcal{N}}^{(2)}(\mathbf{r}_{12}) = \frac{1}{Q_{\mathcal{N}}(\mathcal{N}-2)!} \int d\mathbf{r}_3 \cdots \int d\mathbf{r}_{\mathcal{N}} \exp\left\{-\beta \sum_{i<j} \phi(r_{ij})\right\}, \tag{1.5}$$

where $\rho = \mathcal{N}/V$, $\phi(r_{ij})$ is the potential for the molecule pair (i, j), and $r_{ij} = |\mathbf{r}_i - \mathbf{r}_j|$. The potential energy $U_{\mathcal{N}}(\mathbf{r}_1, \mathbf{r}_2, \cdots, \mathbf{r}_{\mathcal{N}})$ of the system is given by the following equation:

$$U_{\mathcal{N}}(\mathbf{r}_1, \mathbf{r}_2, \cdots, \mathbf{r}_{\mathcal{N}}) = \sum_{i=1}^{\mathcal{N}-1} \sum_{j=i+1}^{\mathcal{N}} \phi(r_{ij}). \tag{1.6}$$

If the radial distribution function $g(r)$ is obtained, the thermodynamic quantities of the fluid can be obtained from the following equations:

$$\rho \kappa kT = 1 + 4\pi\rho \int_0^\infty [g(r) - 1] r^2 dr, \tag{1.7}$$

where,

$$\kappa = -\frac{1}{V}\left(\frac{\partial V}{\partial p}\right)_T = \frac{1}{\rho}\left(\frac{\partial \rho}{\partial p}\right)_T,$$

where, κ is the isothermal compressibility,

$$\frac{E}{\mathcal{N}kT} = \frac{3}{2} + \frac{2\pi\rho}{kT} \int_0^\infty \phi(r) g(r) r^2 dr, \tag{1.8}$$

$$\frac{pV}{\mathcal{N}kT} = 1 - \frac{2\pi\rho}{3kT}\int_0^\infty \frac{d\phi(r)}{dr}g(r)r^2dr. \tag{1.9}$$

Here, E is the energy of the whole system. Equations (1.7), (1.8), and (1.9) are respectively the compressibility equation, the energy equation, and the virial equation. Given $g(r)$ for a given two-molecule interaction $\phi(r)$, all thermodynamic quantities can be derived from these equations. Thus, the fluid physics challenge comes down to finding $g(r)$. Real molecules are often not spherical, but in this book we treat molecules as spherical. Therefore, the intermolecular interaction $\phi(r)$ is a function of only r.

The potential energy u of the system per unit volume is called excess internal energy and is given by the following equation:

$$u = \frac{U_{\mathcal{N}}(\mathbf{r}_1, \mathbf{r}_2, \cdots, \mathbf{r}_{\mathcal{N}})}{V} = 2\pi\rho^2\int_0^\infty dr\, r^2\phi(r)g(r). \tag{1.10}$$

In this case, Eq. (1.8) can be written as follows:

$$E = \frac{3}{2}\mathcal{N}kT + uV. \tag{1.11}$$

1.2 Thermodynamics

1.2.1 Pressure from Energy Equation

If entropy is S, the following equation is obtained from the law of conservation of energy:

$$dS = \frac{1}{T}(dE + pdV - \mu d\mathcal{N}). \tag{1.12}$$

In this section, we assume $\mathcal{N}$ = constant. In this case, the following equation is obtained from Eq. (1.12):

$$\left(\frac{\partial S}{\partial V}\right)_T = \frac{1}{T}\left[\left(\frac{\partial E}{\partial V}\right)_T + p\right], \quad \left(\frac{\partial S}{\partial T}\right)_V = \frac{1}{T}\left(\frac{\partial E}{\partial T}\right)_V. \tag{1.13}$$

The second-order partial derivative of any thermodynamic function f with respect to any two independent thermodynamic variables x and y is independent of the order of differentiation, so $(\partial^2 f/\partial x\partial y) = (\partial^2 f/\partial y\partial x)$ holds. Therefore, the following equation is obtained from Eq. (1.13):

$$\left(\frac{\partial^2 S}{\partial T \partial V}\right) = -\frac{1}{T^2}\left[\left(\frac{\partial E}{\partial V}\right)_T + p\right] + \frac{1}{T}\left[\frac{\partial^2 E}{\partial T \partial V} + \left(\frac{\partial p}{\partial T}\right)_V\right]$$

$$= \frac{1}{T}\frac{\partial^2 E}{\partial V \partial T}.$$

$$\therefore \left(\frac{\partial p}{\partial T}\right)_V = \frac{1}{T}\left[\left(\frac{\partial E}{\partial V}\right)_T + p\right]. \tag{1.14}$$

From Eq. (1.11), the following equation is obtained:

$$\left(\frac{\partial E}{\partial V}\right)_T = u + V\left(\frac{\partial u}{\partial V}\right)_T = u - \rho\left(\frac{\partial u}{\partial \rho}\right)_T. \tag{1.15}$$

From Eqs. (1.14) and (1.15), we obtain the following equation:

$$\alpha \equiv \left(\frac{\partial p}{\partial kT}\right)_\rho = \frac{1}{kT}\left[u - \rho\left(\frac{\partial u}{\partial \rho}\right)_T + p\right]. \tag{1.16}$$

Rewrite Eq. (1.16) to get

$$\left(\frac{\partial \beta p}{\partial \beta}\right)_\rho = -u + \rho\left(\frac{\partial u}{\partial \rho}\right)_\beta. \tag{1.17}$$

If Eq. (1.16) is integrated with kT or Eq. (1.17) is integrated with β, the pressure p is obtained.

1.2.2 Chemical Potential from Energy Equation

In this section, we assume V = constant. In this case, the following equation is obtained from Eq. (1.12):

$$\left(\frac{\partial S}{\partial \mathcal{N}}\right)_T = \frac{1}{T}\left[\left(\frac{\partial E}{\partial \mathcal{N}}\right)_T - \mu\right], \left(\frac{\partial S}{\partial T}\right)_\mathcal{N} = \frac{1}{T}\left(\frac{\partial E}{\partial T}\right)_\mathcal{N}. \tag{1.18}$$

Using the same method as Section 1.2.1, the following expression is obtained from Eq. (1.18):

$$\left(\frac{\partial^2 S}{\partial T \partial \mathcal{N}}\right) = -\frac{1}{T^2}\left[\left(\frac{\partial E}{\partial \mathcal{N}}\right)_T - \mu\right] + \frac{1}{T}\left[\left(\frac{\partial^2 E}{\partial T \partial \mathcal{N}}\right) - \left(\frac{\partial \mu}{\partial T}\right)_\mathcal{N}\right]$$

$$= \frac{1}{T}\left(\frac{\partial^2 E}{\partial \mathcal{N} \partial T}\right).$$

$$\therefore \left(\frac{\partial \mu}{\partial T}\right)_\mathcal{N} = -\frac{1}{T}\left[\left(\frac{\partial E}{\partial \mathcal{N}}\right)_T - \mu\right] = -\frac{1}{T}\left[\frac{3}{2}kT + \left(\frac{\partial u}{\partial \rho}\right)_T - \mu\right].$$

$$\therefore \left(\frac{\partial\beta\mu}{\partial\beta}\right)_\rho = \mu + \beta\left(\frac{\partial\mu}{\partial\beta}\right)_\rho = \mu - T\left(\frac{\partial\mu}{\partial T}\right)_\rho = \mu + \frac{3}{2}kT + \left(\frac{\partial u}{\partial\rho}\right)_T - \mu = \frac{3}{2}kT + \left(\frac{\partial u}{\partial\rho}\right)_\beta . \tag{1.19}$$

$$\therefore \left(\frac{\partial\beta\mu}{\partial\beta}\right)_\rho = \frac{3}{2\beta} + \left(\frac{\partial u}{\partial\rho}\right)_\beta . \tag{1.20}$$

Integrating both sides of Eq. (1.20) with β gives the chemical potential μ (see Section 7.4).

1.2.3 Pressure and Chemical Potential from Compressibility Equation

From the second expression of Eq. (1.7), the following expression is obtained:

$$\left(\frac{\partial\beta p}{\partial\rho}\right)_\beta = \frac{\beta}{\kappa\rho} \equiv \frac{1}{\chi_{\mathrm{red}}}, \tag{1.21}$$

where, χ_{red} is reduced isothermal compressibility. From Gibbs–Duhem equation

$$SdT - Vdp + \mathcal{N}d\mu = 0. \tag{1.22}$$

If T = constant, $dp = \rho d\mu$ from Eq. (1.22), so the following expression is obtained from Eq. (1.21):

$$\left(\frac{\partial\beta\mu}{\partial\rho}\right)_\beta = \frac{1}{\rho}\left(\frac{\partial\beta p}{\partial\rho}\right)_\beta = \frac{1}{\rho\chi_{\mathrm{red}}}. \tag{1.23}$$

Integrating Eq. (1.23) over ρ gives p and μ.

1.3 Consistency between Energy Equation and Compressibility Equation

If the theory of thermodynamics and statistical mechanics is strictly correct, the same thermodynamic quantities should match regardless of their derivation paths. Here, we obtain the conditional expression for a physical quantity obtained from the energy equation coinciding with that from the compressibility equation.

Apply $\rho(\partial/\partial\rho)_\beta$ to both sides of Eq. (1.20), change the order of differentiation, and Eq. (1.23) gives

$$\rho\left(\frac{\partial}{\partial\rho}\right)_\beta\left(\frac{\partial\beta\mu}{\partial\beta}\right)_\rho=\left(\frac{\partial}{\partial\beta}\right)_\rho\rho\left(\frac{\partial\beta\mu}{\partial\rho}\right)_\beta=\left(\frac{\partial}{\partial\beta}\frac{1}{\chi_{\text{red}}}\right)_\rho$$
$$=\rho\left(\frac{\partial^2 u}{\partial\rho^2}\right)_\beta.$$

$$\therefore\ \frac{\partial}{\partial\beta}\left(\frac{1}{\chi_{\text{red}}}\right)=\rho\frac{\partial^2 u}{\partial\rho^2}. \tag{1.24}$$

This partial differential equation is a conditional expression to be obtained. This is one of the basic equations of Self-Consistent Ornstein–Zernike Approximation (SCOZA) described in Chapter 5.

Chapter 2

Strange Temperature Change of Water Density

Ordinary substances expand when heated and contract when cooled (positive thermal expansion). However, in the case of water, positive thermal expansion occurs when it is above 4°C, and below 4°C it shrinks when heated and expands when cooled (negative thermal expansion). This mysterious nature of water has been investigated for hundreds of years, and various theories have been proposed by many researchers to solve the mystery. However, none of these theories has solved this. The author has recently fundamentally elucidated the thermodynamic mechanism behind negative thermal expansion. It is discussed in detail in this chapter.

2.1 Positive Thermal Expansion

Attractive forces act between the water molecules, which work to increase the density by condensing the water molecule population. On the other hand, each water molecule has a thermal disorder motion, and the energy of the disorder motion increases as it warms up, so that it expands against the intermolecular attractive forces.

The Physics of Liquid Water
Makoto Yasutomi

ISBN 978-981-4877-25-1 (Hardcover), 978-1-003-05616-4 (eBook)
www.jennystanford.com

However, when temperature and density are reached such that both shrinkage and expansion are balanced, a thermal equilibrium state is maintained. This is the thermodynamic mechanism of positive thermal expansion. So what is the thermodynamic mechanism that produces negative thermal expansion? It is one of the main purposes of this book to clarify this thermodynamic mechanism.

2.2 Negative Thermal Expansion

In Fig. 2.1, the experimental data (closed circles) [Abascal and Vega (2005); Lide (1990)] show that the density of water ρ changes with temperature T under one atmospheric pressure. Above 4°C, the density decreases as the temperature rises (positive thermal expansion), but below 4°C the density increases with temperature (negative thermal expansion). Analyzing this experimental fact based on the laws of thermodynamics, we can get an important conclusion about the thermodynamic mechanism of negative thermal expansion, as explained below.

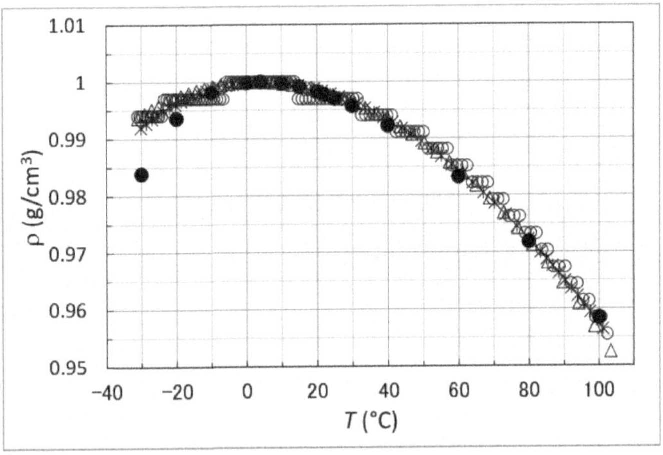

Figure 2.1 Density-temperature relation of liquid water at 1 bar. Closed circles, asterisks, triangles, and open circles show those for experimentally measured data [Abascal and Vega (2005); Lide (1990)], ϕ_{c1}, ϕ_{c2}, and ϕ_{c3}, respectively, where $\phi_{ci} = \phi_i$ in Ref. [Yasutomi (2016)].

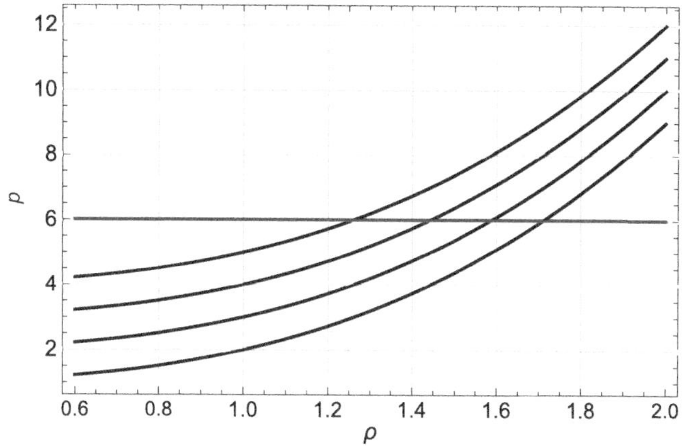

Figure 2.2 Schematic pressure isothermal curves (black lines): The intersections of the constant pressure line (red line) and the pressure isothermal curves give the relationship between density and temperature under constant pressure. When the temperature is raised above 4°C, the isothermal curve of the pressure moves to the high pressure side ($\alpha > 0$), and positive thermal expansion occurs. Below 4°C, the reverse behavior ($\alpha < 0$) occurs, so negative thermal expansion occurs.

A curve representing the relationship between pressure p and density ρ, when the temperature is kept constant, is called a pressure isothermal curve. For any substance, if the temperature is kept constant, the pressure increases with density, so the isothermal curve becomes a monotonically increasing curve (black lines in the Fig. 2.2). For normal substances and water above 4°C, if the density is constant, the pressure p increases with the temperature T, so $\alpha = (\partial p/\partial kT)_\rho > 0$. Therefore, the isothermal curve moves to the high pressure side (from bottom to top in the Fig. 2.2) as the temperature rises. In the Fig. 2.2, the intersections of the constant pressure horizontal line (red line) and the isothermal curves (black lines) give the relationship between density and temperature under constant pressure (isobaric curve). Of course, it can be seen that the density decreases with increasing temperature (positive thermal expansion). In other words, positive thermal expansion is equivalent to the pressure isothermal curve moving to the high pressure side as the temperature rises, and it can be seen that this phenomenon occurs when $\alpha > 0$.

From similar considerations, we can see that the negative thermal expansion occurs when $\alpha < 0$ below 4°C. Therefore, the basic elucidation of the mystery of negative thermal expansion is to clarify the thermodynamic mechanism that switches the sign of α from positive to negative at 4°C with cooling in relation to the functional form of the interaction between water molecules.

Chapter 3

Fundamental Clarification of Thermodynamic Phenomena in Water

The thermodynamic properties of a material are determined by the interactions between the particles that make the material up. These can be derived using thermodynamic laws and statistical mechanics techniques. Therefore, if the relationship between the functional form of the interparticle interactions and the thermodynamic phenomenon can be established using thermodynamic quantities, it can be said that the mystery of the phenomenon has been fundamentally solved.

Thermodynamic quantities can be derived from various thermodynamic potentials, but in this chapter we use excess internal energy $u(\rho, \beta)$ defined by Eq. (1.10). The reason is that u has a clear relationship with the functional form of the interaction between particles, as can be seen from Eq. (1.10). In this sense, u is the most suitable for fundamentally elucidating the thermodynamic mechanism that generates negative thermal expansion compared to other thermodynamic potentials.

As can be seen from Eq. (1.16), the sign of α is determined by the competition among p, u, and $\rho(\partial u/\partial \rho)_T$. Since p is always positive, it

The Physics of Liquid Water
Makoto Yasutomi

ISBN 978-981-4877-25-1 (Hardcover), 978-1-003-05616-4 (eBook)
www.jennystanford.com

works to make α positive. However, u and $\rho(\partial u/\partial \rho)_T$ can be positive or negative depending on the thermodynamic state of the liquid. Therefore, it can be seen that these two terms play an important role in switching the sign of α. We need to elucidate these thermal behaviors in relation to the shape of the interparticle interaction $\phi(r)$.

3.1 Function Form of Interaction between Water Molecules

Since the thermodynamic properties of matter are determined by the functional form of the intermolecular interaction $\phi(r)$, this section considers how to determine the functional form.

3.1.1 Water Molecule Shape

Natural water includes nine isotopes of water molecules of $H_2^{16}O$, $H_2^{18}O$, $H_2^{17}O$, $HD^{16}O$, $HD^{18}O$, $HD^{17}O$, $D_2^{16}O$, $D_2^{18}O$, and $D_2^{17}O$. Out of them, $H_2^{16}O$ accounts for 99.76%. In this book, this $H_2^{16}O$ is simply written as H_2O, and a single component fluid composed of this molecule alone is considered.

The shape of the water molecule H_2O is shown in Fig. 3.1. It is almost a sphere with a radius of 1.5 Å. The O-H bond distance of H_2O obtained from experiments by [Benedict (1956)] is $r_{OH} = 0.9572$ Å, and the H-O-H bond angle is $\theta_{HOH} = 104.523^\circ$. The van der Waals radii of O and H defined by [Pauling (1935)] are 1.4 Å and 1.2 Å, respectively. It is known that the shape of a molecule changes slightly depending on the quantum state of the molecule. It has been reported that the H-O-H bond angle and the O-H bond distance change slightly when excited to a high rotational level or excited to a high electronic state. However, the vibrational state does not seem to significantly affect the size of the molecule. These matters should be taken into account in order to determine the functional form of the interaction between water molecules in the water. There are three possible ways to accomplish this.

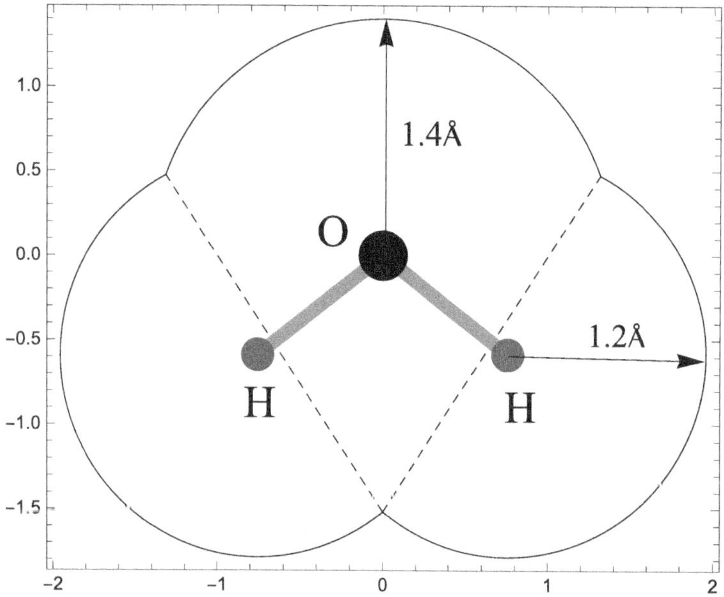

Figure 3.1 Water molecule shape. O: Center of oxygen atom. H: Center of hydrogen atom. Large arc: Part of the van der Waals sphere with a radius of 1.4 Å centered on oxygen atoms. Two slightly smaller arcs: Part of van der Waals spheres with a radius of 1.2 Å centered on hydrogen atom.

3.2 Method by Solving the Basic Equation of Quantum Mechanics

The nature of water molecules can be derived from the solution of the fundamental equations of quantum mechanics. One water molecule consists of 13 particles of one oxygen nucleus, two hydrogen nuclei, and 10 electrons. The behavior of these 13 particles can be determined if, for example, the wave function is obtained by solving the Schrödinger equation. All required physical quantities can be derived from this wave function. However, it is not so easy to solve quantum mechanics equations and determine the properties of even one water molecule.

Next, consider deriving the interaction between two water molecules from quantum mechanics equations. In order to find the interaction between two molecules, it is necessary to find the wave function by solving the equations for a system consisting of a total of 26 particles. This is even more difficult than the problem of one water molecule. In actual water, there are many other molecules around the two molecules we are thinking about, so we have to consider the effects from these molecules. Therefore, it can be said that it is almost impossible to obtain the interaction between water molecules by such a method. Usually, the following two methods are frequently used.

3.3 Method by Using Realistic Water Model

One way is to use a realistic model. These models include TIP3P, TIP4P [Jorgensen (1983)], SPC/E [Berendsen (1987)], TIP5P [Mahoney and Jorgensen (2000)], TIP4P/2005 [Abascal and Vega (2005)], mW [Molinero and Moore (2009)], etc. In the realistic model, the charge distribution in the water molecule is divided into several positive point charges q_{H} and negative point charges q_{G} ($= -q_{\mathrm{H}}$). And each point charge is considered to be fixed and arranged at a fixed position with respect to the neutral oxygen atom. As an example, the water molecule model of TIP5P is shown in Fig. 3.2. Each parameter is shown in Table 3.1. The symbol e is the amount of proton electricity, r_{OL} is the distance between the oxygen atom and the negative point charge L, and θ_{LOL} is the angle formed by (L-O-L). The Lennard-Jones potential acts between oxygen atoms, and the Coulomb force acts between each charge. In this case, the

Table 3.1 Parameters in TIPnP [Mahoney and Jorgensen (2000)]

	TIP3P	TIP4P	TIP5P
$q_{\mathrm{H}}(e)$	0.417	0.520	0.241
σ_{O}(Å)	3.15061	3.15365	3.12
ϵ_{O}(kcal/mol)	0.1521	0.1550	0.16
r_{OH}(Å)	0.9572	0.9572	0.9572
θ_{HOH}(deg)	104.52	104.52	104.52
r_{OL}(Å)		0.15	0.70
θ_{LOL}(deg)			109.47

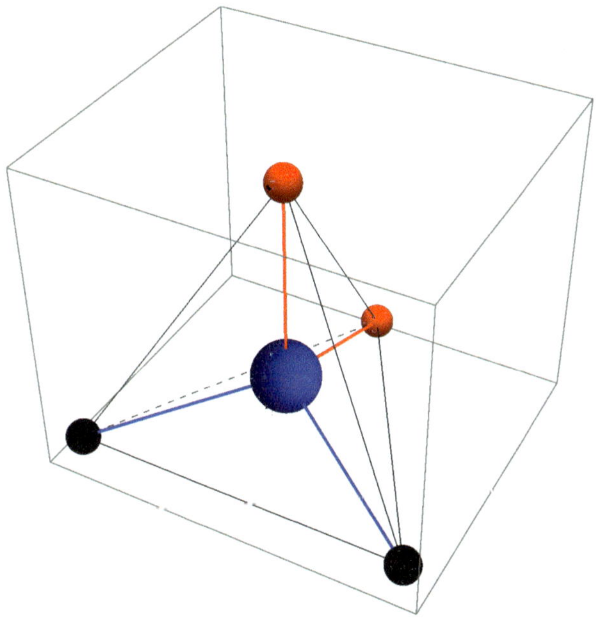

Figure 3.2 TIP5P water molecule model. Blue sphere: Neutral oxygen atom. Black sphere: Positive point charge. Red sphere: Negative point charge.

interaction ϕ_{ab} between water molecules a and b is given by the following equation:

$$\phi_{\text{ab}} = \sum_{ij} \frac{q_i q_j e^2}{r_{ij}} + 4\epsilon_0 \left[\left(\frac{\sigma_0}{r_{\text{OO}}} \right)^{12} - \left(\frac{\sigma_0}{r_{\text{OO}}} \right)^{6} \right]. \qquad (3.1)$$

3.4 Simplified Model (Core-Softened Model)

The other way is to use a core-softened model that is built primarily to elucidate the physics behind the thermodynamic phenomenon of interest. In this model, water molecules are regarded as spheres, and the interaction between particles is approximated by hard-core (HC) repulsion and tails. There are three types of tails: soft-repulsion, attraction, and soft-repulsion plus attraction. The thermodynamic behavior of a finite number of such water molecule populations are investigated by using the Monte Carlo method or molecular dynamics method [Caballero and Puertas (2006); Fomin (2011); Gibson and Wilding (2006); Gribova (2009); Jagla (2001); Lomba (2007); Malescio (2011); Oliveira (2009); Prestipino (2010); Sadr (1998); Wilding and Magee (2002); Yan (2004, 2008)]. Tails of various shapes have been proposed and succeeded in qualitatively reproducing negative thermal expansion. However, no quantitative comparison with experimental data has been made, and it cannot be said that the physics behind the negative thermal expansion has been elucidated. In the simulation of a finite number of particle populations, sometimes the particle distribution varies, and it is observed that a mixed state of a high density region and a low density region is realized. Although there is an idea that density anomaly is generated by changing the abundance ratio of these two regions, there is no explanation about the mechanism in which the proportion of the low density region increases when it is cooled below 4°C.

3.5 Method by Self-Consistent Ornstein–Zernike Approximation

To elucidate what kind of thermodynamic mechanism generates the density anomaly of liquid water, it is necessary to build a model in which only the physical properties essential for the phenomenon are taken in, and other miscellaneous properties are excluded. It is also important to reproduce the experimental results. At present, the author believes that the most suitable model for this purpose is the thermodynamic self-consistent theory (SCOZA). Since SCOZA uses a method to directly determine the thermodynamic potential $u(\rho, \beta)$, as described in Chapter 5, it is easy to discuss in relation to the functional form of interaction. Therefore, it can be said that this is the optimal theory for fundamentally elucidating thermodynamic phenomena.

In this book, water molecules are approximated by spherical particles, and an interaction consisting of a hard-core (HC) and tails work between the molecules. And we will investigate the thermodynamic properties of water using SCOZA. The potential tails are given by Yukawa terms. In this case, as will be described later, analytical solutions of the Ornstein–Zernike (OZ) equations for the MSA can be used (see Chapter 5). Compared with the Monte Carlo method and molecular dynamics method, this method can process a large amount of calculations efficiently and in a short time. Further, there is an advantage that the calculation results can be easily analyzed in association with the function form of the interparticle interaction. In addition, since any smooth function can be approximated with high accuracy in Yukawa terms, this method can be applied to interactions of various shapes. In addition, the uniformity of the particle distribution is maintained, and there is no problem associated with the variation in density distribution observed in the Monte Carlo method and molecular dynamics method using a finite number of particle populations. Therefore, it is suitable for elucidating the physics behind negative thermal expansion. SCOZA is entirely self-contained, and the only necessary input physics is a function that expresses the interaction between particles. It is also well known as a theory that can reproduce the thermodynamic properties of fluids with high accuracy [Betancourt-Cárdenas (2008); Pini (1998a,b); Schöll-Paschinger (2004, 2005); Yasutomi (2010, 2014, 2015, 2016)]. We also clarify why numerical simulations cannot determine the vaporization and liquefaction points in the narrow-temperature region just below the critical temperature.

3.6 Multi-Yukawa Type Intermolecular Interaction

In SCOZA, the interaction between molecules consists of hard-core potential and tail potential. If multi-Yukawa or screened power-series potential [Yasutomi (2010)] is adopted as the tail potential, there is an advantage that the analytical solution obtained for MSA can be used. In this book, we use Yukawa type given by the following formula:

$$\phi(r) = \begin{cases} \infty, & r < 1, \\ -\sum_{n=2}^{N} a_n \dfrac{\exp[-z_n(r-1)]}{r}, & r \geq 1. \end{cases} \tag{3.2}$$

Here, N represents an arbitrary integer, and a_n and z_n are arbitrary constants. Since an arbitrarily smooth potential can be developed with multi-Yukawa terms, this SCOZA can be applied to fluids with various interparticle interactions.

SCOZA will be explained in detail in Section 5.6. In this book, we focus on the negative thermal expansion, the diversity of water and ice structures, and the temperature dependence of the isothermal compressibility among many strange thermodynamic properties of water. And we will consider the physics behind these phenomena in relation to the shape of the intermolecular interaction [Yasutomi (2014, 2015, 2016)].

3.7 Thermodynamic Mechanism of Negative Thermal Expansion

To reveal the physics behind the mysterious temperature dependence of water density, we consider two typical intermolecular interactions $\phi_{\mathrm{b}2}$ and $\phi_{\mathrm{b}5}$ (see Table 3.2). The former produces negative thermal expansion but the latter doesn't. And we compare the phenomena caused by these interactions. As shown in Fig. 3.3, $\phi_{\mathrm{b}2}$ consists of HC repulsion and soft-repulsive tail followed by attractive tail. On the other hand, $\phi_{\mathrm{b}5}$ consists of HC repulsion and attractive tail, and does not have soft-repulsive tail.

Though $\phi_{\mathrm{b}6}$ (see Table 3.2) is not shown in Fig. 3.3, $\phi_{\mathrm{b}2}$ and $\phi_{\mathrm{b}6}$ have similar function shape and produce qualitatively the same thermodynamic phenomena. Therefore, the excess internal energy u generated by $\phi_{\mathrm{b}6}$ is shown in Fig. 3.4 as a function of ρ for three temperatures T (°C) $= +20, +4, -20$ by dotted lines. u moves to the high energy side as the temperature decreases. As a result, as shown in Fig. 3.5, α moves from positive to negative near $\rho = 0.3$ as

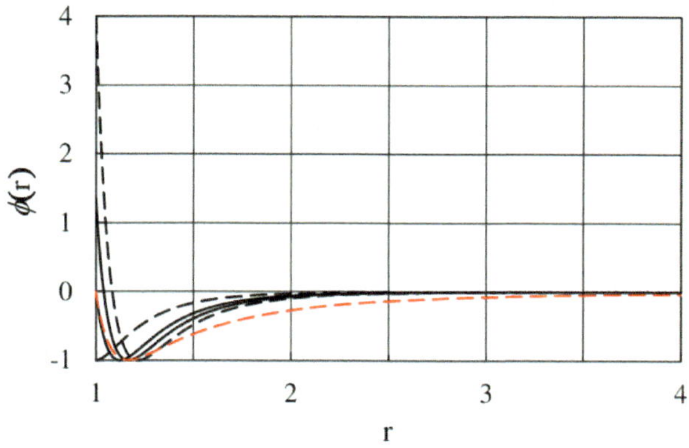

Figure 3.3 Potential tails causing a density anomaly. The full lines represent potentials inducing a density anomaly: ϕ_{b2} and ϕ_{b3} (from the top to the bottom at the HC contact in the figure, respectively). The dashed lines represent potentials never inducing negative thermal expansion: ϕ_{b1}, ϕ_{b7} (red line), and ϕ_{b5} (from the top to the bottom at the HC contact, respectively), where $\phi_{bi} = \phi_i$ in Ref. [Yasutomi (2015)]. See Table 3.2.

Table 3.2 Parameters for ϕ_{b1}, ϕ_{b2}, ϕ_{b3}, ϕ_{b5}, ϕ_{b6}, and ϕ_{b7}

	ϕ_{b1}			ϕ_{b2}			ϕ_{b3}	
n	z_n	a_n	n	z_n	a_n	n	z_n	a_n
2	2.65	2.73132	2	1.184	0.16399	2	1.1201	0.147233
3	6.636	0.540524	3	3.192	2.5134	3	3.3921	2.18259
4	12.788	−7.2524	4	13.204	−4.0191	4	13.5	−2.32906
	ϕ_{b5}			ϕ_{b6}			ϕ_{b7}	
n	z_n	a_n	n	z_n	a_n	n	z_n	a_n
2	1.7024	0.23543	2	0.54052	4.25009	2	0.35	0.3547
3	4.1424	1.2028	3	3.9274	−25.1902	3	1.45	1.2141
4	14.564	−0.43837	4	7.73812	−7.9332	4	2.508	0.2082
						5	9.84	−1.0519
						6	14.974	−0.7186

the temperature decreases. It is explained below that this movement creates a pressure anomally and a density anomally.

As can be seen from Eq. (1.16), p is always positive, so it has the property of making α positive. On the other hand, u can be either positive or negative depending on $\phi(1^+)$, as can be seen from Eq. (1.10) (here, $1^+ = \lim_{\delta\to 0}(1 + |\delta|)$). At high density and

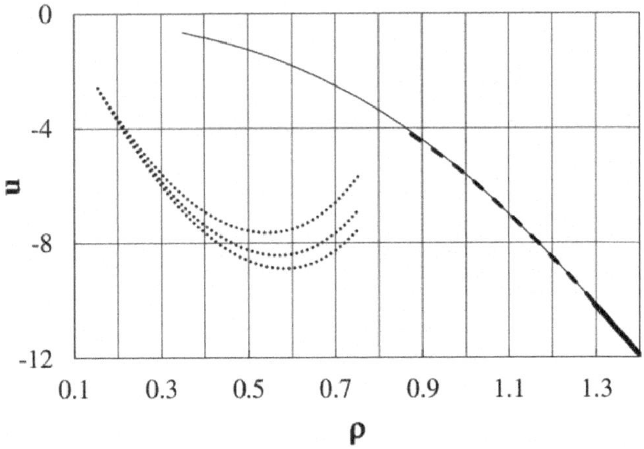

Figure 3.4 The excess internal energy u per unit volume as a function of density ρ. The dotted lines show the u–ρ curves at T (°C) = +20, +4, and −20 (from the bottom to the top of the figure, respectively) for ϕ_{b6}. The thin full, dashed, and thick full lines are for kT = 0.896, 0.372, and 0.0324, respectively, for ϕ_{b5}. See Ref. [Yasutomi (2015)].

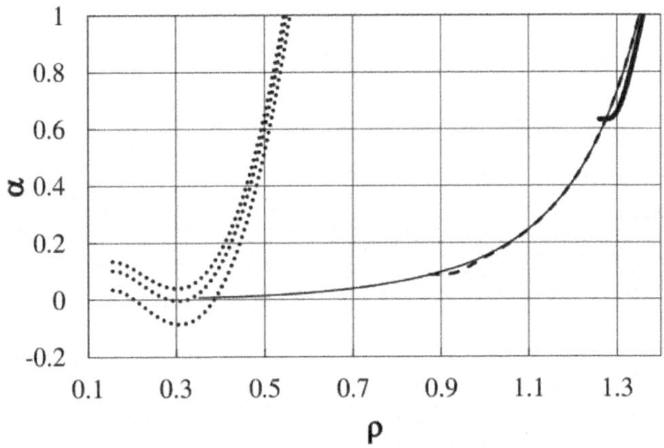

Figure 3.5 The differential coefficient $\alpha = \partial p / \partial (kT)_\rho$ as a function of density. The dotted lines are at T (°C) = +20, +4, and −20 (from the top to the bottom of the figure, respectively) for ϕ_{b6}. The thin full, dashed, and thick full lines represent $\alpha/100$ for kT = 0.896, 0.372, and 0.0324, respectively, for ϕ_{b5}. See Ref. [Yasutomi (2015)].

low temperature, the distribution function $g(r)$ has a positive large value near the HC contact ($r = 1$). Therefore, the sign of u is determined depending on the sign of $\phi(1^+)$, and the behavior of u determines $\rho(\partial u/\partial \rho)_T$. In this way, the behavior of $\phi(1^+)$ and u plays an important role in determining the sign of α.

To investigate the relationship between the shape of $\phi(r)$ and the thermodynamic behavior of u in more detail, the integrand included in Eq. (1.10) is set to $f(r, \rho, T) = r^2\phi(r)g(r)$ and introduce the following function $\Delta f(r, \rho, T)$:

$$\Delta f(r, \rho, T) = f(r, \rho, T) - f(r, \rho, +20^\circ\text{C}). \tag{3.3}$$

In the above equation, f is written as a function of r, ρ, and T, because $g(r)$ is also a function of ρ and T.

The function $\Delta f(r, \rho, T)$ is shown in Fig. 3.6 as a function of r. In the figure, the solid line is $\Delta f(r, \rho^*, +4^\circ\text{C})$, and the dashed line is $\Delta f(r, \rho^*, -20^\circ\text{C})$. Here, $\rho^* = 0.7535$. As shown in the figure, any $\Delta f(r, \rho, T)$ is small in the region away from the hard sphere, but extremely large near the HC contact, and it has increased rapidly with the temperature descent. This is true for other densities than $\rho^* = 0.7535$. Therefore, it can be seen that the value of u and its behavior are mainly determined near the hard sphere contact where $\phi(r)$ acts as a large repulsive force. This increases the value of

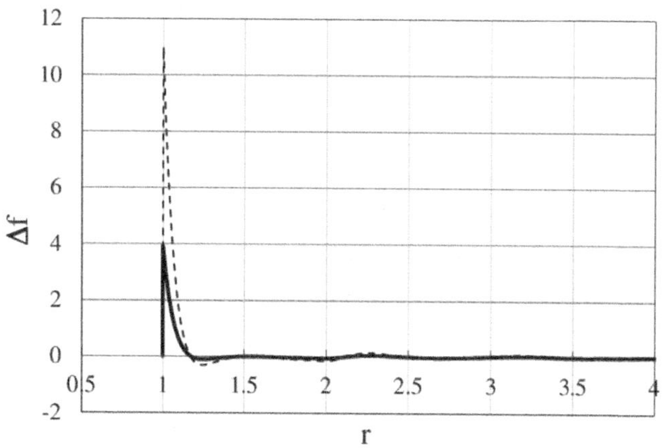

Figure 3.6 The difference $\Delta f(r, \rho, T)$ as a function of distance r. The full and dashed lines show $\Delta f(r, \rho^*, +4^\circ\text{C})$ and $\Delta f(r, \rho^*, -20^\circ\text{C})$, respectively, where $\rho^* = 0.7535$. See Ref. [Yasutomi (2015)].

$\rho(\partial u/\partial \rho)_T$ and, as a result, makes the value of α negative, producing negative thermal expansion. For ϕ_{b2}, a similar thermodynamic mechanism has been also confirmed to generate negative thermal expansion.

For comparison, u is shown as a function of ρ in Fig. 3.4 and α in Fig. 3.5 for an interaction ϕ_{b5} that does not generate density anomaly. In these figures, the thin solid line, the broken line, and the thick solid line represent functions for $kT = 0.896$, 0.372, and 0.0324, respectively. ϕ_{b5} has no soft-repulsive tail, so $\phi(1^+)$ has a negative minimum. This leads to a negative minimum for u and $\rho(\partial u/\partial \rho)_T$ at high density and low temperature, and as a result, α is always kept positive, and it prevents the negative thermal expansion from occurring.

3.7.1 Shape of Intermolecular Interaction That Causes Negative Thermal Expansion

In order to investigate what shape of the intermolecular interaction causes negative thermal expansion, five typical potential-tails are shown in Fig. 3.3. Two functions $\phi_{b2}(r)$ and $\phi_{b3}(r)$, $[\phi_{b2}(1^+) > \phi_{b3}(1^+)]$, shown as solid lines in the figure cause negative thermal expansion, but the three functions $\phi_{b1}(r)$, $\phi_{b7}(r)$, and $\phi_{b5}(r)$, $[\phi_{b1}(1^+) > \phi_{b7}(1^+) > \phi_{b5}(1^+)]$, shown in dashed lines do not cause negative thermal expansion. That is, if $\phi(1^+)$ is within a certain range, negative thermal expansion is induced, but if $\phi(1^+)$ is too large or too small, negative thermal expansion is not generated. This range also depends on the shape of the attractive-tail. In fact, ϕ_{b3} and ϕ_{b7} have almost the same soft-repulsive tail shape, but the attractive-tail shape is very different. The former induces positive and negative thermal expansion depending on temperature, while the latter produces only positive thermal expansion at any temperature.

Chapter 4

Variety of Shapes of Water Molecule Interactions

There are many thermodynamic potentials and various thermodynamic quantities can be derived from any of them. The excess internal energy $u(\rho, \beta)$ is one of the thermodynamic potentials. As can be seen from Eq. (1.10), it consists of the product of the intermolecular interaction and the distribution function $\phi(r)g(r)$. Therefore, if $\phi(r)g(r)$ is the same, the same thermodynamic properties can be obtained. There can be an infinite number of combinations that give the same $\phi(r)g(r)$, so there can be an infinite number of $\phi(r)$ that produce the same thermodynamic phenomenon. In fact, the author has discovered a number of intermolecular interactions $\phi(r)$ that reproduce the temperature dependence of water density under atmospheric pressure with sufficient accuracy. Some of them ϕ_{c1}, ϕ_{c2}, and ϕ_{c3} are shown in Fig. 4.1. The parameters a_n included in the potential tail are shown in Table 4.1. Regarding z_n, it is assumed that $z_n = 0.6395(n-1)$, $1.18(n-1)$, and $0.6[1+8(n-2)]$, respectively, for ϕ_{c1}, ϕ_{c2}, and ϕ_{c3}. The density anomaly produced by these potential tails are shown in Fig. 2.1, and the experimental results are reproduced with sufficient accuracy. The shapes of these potential tails are

The Physics of Liquid Water
Makoto Yasutomi

ISBN 978-981-4877-25-1 (Hardcover), 978-1-003-05616-4 (eBook)
www.jennystanford.com

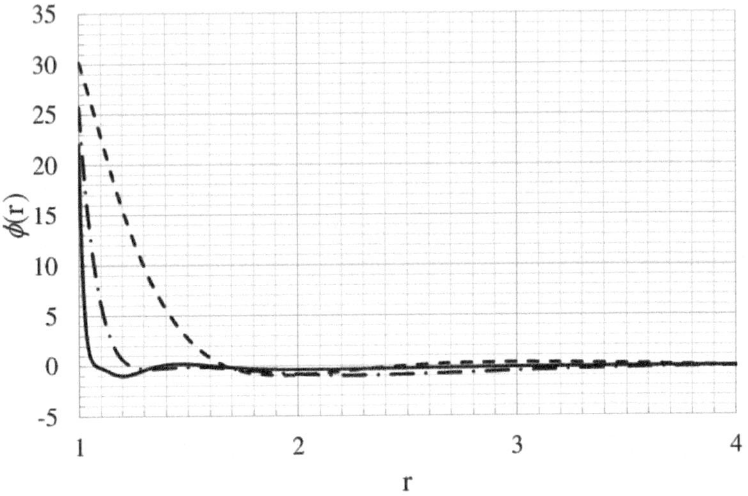

Figure 4.1 Potential tails. Dashed, chain-dashed, and full lines represent ϕ_{c1}, ϕ_{c2}, and ϕ_{c3}, respectively, where $\phi_{ci} = \phi_i$ in Ref. [Yasutomi (2016)].

Table 4.1 Parameters a_n for ϕ_{c1}, ϕ_{c2}, and ϕ_{c3}

n	ϕ_{c1} a_n	ϕ_{c2} a_n	ϕ_{c3} a_n
2	+39.5839	+73.1654	+1.86973
3	−433.906	−320.979	−53.9562
4	+1478.56	−7597.99	+415.834
5	−1906.08	+64001.9	−1189.08
6	+791.644	−134345.	+1456.07
7			−652.705

very different from the shapes of ϕ_{b2} and ϕ_{b3} shown in Fig. 3.3. The thermodynamic mechanism that produces density anomaly is qualitatively identical. The discovery that there can be a variety of water molecule interactions that produce the same thermodynamic phenomenon will play a very important role in unraveling the mysteries of water and ice structural diversity [Eisenberg and Kauzmann (1969)].

4.1 Soft-Repulsive Tail Slope and Isothermal Compression Ratio

We mentioned above the density anomaly and the diversity of water and ice structures as mysterious properties of water that are not found in ordinary substances. The temperature dependence of the isothermal compressibility of water is also complex, and full of mysteries. To elucidate this mystery, it is necessary to clarify the relationship between the isothermal compressibility and the shape of the intermolecular interaction $\phi(r)$. We show the temperature dependences of the isothermal compressibilites in Fig. 4.2. Those are obtained for potential-tail ϕ_{c1}, ϕ_{c2}, and ϕ_{c3} shown in Fig. 4.1. As can be seen from these figures, the steeper the soft-repulsive potential gradient, the smaller the isothermal compression ratio. This is an intuitive conclusion. Since the shape of the water molecule is deformed by the quantum state of the molecule, it is quite possible that it affects the gradient of the soft-repulsive potential. The gradient should also depend on the coordination and orientation between molecules. Since the distribution of coordination and orientation should change depending on temperature

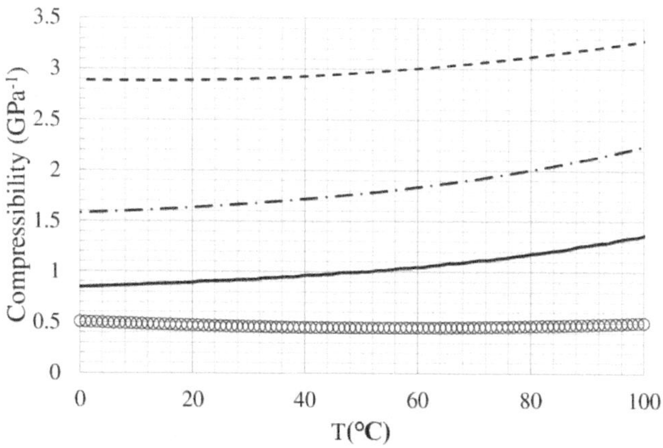

Figure 4.2 Isothermal compressibility–temperature relation for liquid water at 1 bar. The dashed line, chain-dashed line, solid line, and open circles represent the results obtained for ϕ_{c1}, ϕ_{c2}, ϕ_{c3}, and experimentally measured data, respectively. See Ref. [Yasutomi (2016)].

and density, it is considered that such influence should be taken into consideration. It is presumed that these complications make the temperature dependence of the isothermal compression rate complicated.

So far, in this book, we have elucidated the thermodynamic mechanism which induces the density anomaly of water in relation to the shape of the intermolecular interactions, that the isothermal compressibility of water is determined by the gradient of the soft-repulsive potential tail, and that the innumerable intermolecular interactions cause the same thermodynamic phenomenon. However, none of the other theories that have been proposed so far has discussed the mysteries in relation to the shape of intermolecular interactions. Therefore, in the following sections, we review some typical theories among them, and discuss the problems involved.

4.2 Problems with Existing Ideas on Negative Thermal Expansion

Reference [Röntgen (1892)] proposed a theory that water consists of two mixtures of high- and low-density components to explain the strange thermodynamic behavior of water. The low-density component is in a low energy state like ice, the high-density component is in a high energy state, and the reason why the density of water is maximum at 4°C is explained as follows.

Above 4°C, the water density increases because the proportion of high-density components increases and the proportion of low-density components decreases as the water cools (positive thermal expansion). But on the other hand, when cooling below 4°C, the ratio of the high-density components decreases, the ratio of the low-density components increases, and the water density decreases (negative thermal expansion). However, there is no explanation why the change in the mixing ratio of the two components is reversed at 4°C as the temperature changes.

As we have explained in Section 2.2, it is clear from the simple and clear law of thermodynamics and the experimental facts that the density anomaly is caused by the inversion of the sign of $\alpha = (\partial p/\partial kT)_\rho$. The inversion occurs at 4°C when the temperature increases or decreases.

In Section 3.7, the author elucidated this mechanism in relation to the functional form of intermolecular interactions. In the binary mixture model of Ref. [Röntgen (1892)], there is no discussion related to the functional shape of intermolecular interaction as elucidated by this author.

Several ideas similar to the binary mixture model described above have been proposed. Reference [Russo and Tanaka (2014)] proposed a water model with two states, S-state and ρ-state, to explain the thermodynamic properties of water. S-state is in a state of low energy, low density, and low degeneracy, but ρ-state is in a state of high degeneration, high energy, and high density. They reproduced the water density anomaly experimental results using molecular dynamics and showed that it was brought about by the change of the abundance ratio of S-state and ρ-state due to the temperature change. However, the relationship between the temperature change of the abundance ratios of the two states and the functional form of the intermolecular interaction has not been clarified. That is, below 4°C, as the temperature decreases, negative thermal expansion is induced as a result of the S-state existence ratio exceeding the ρ-state existence ratio. But the reason for it remains a mystery. This mystery can be easily explained as follows based on the thermodynamic mechanism elucidated by the author in Section 2.2 and Section 3.7.

If the water temperature is higher than 4°C, the thermal random motion energy of each molecule decreases when water is cooled, so the motion energy to expand is reduced and the water is condensed due to the intermolecular attractive force. It results in the density increase (the increase of existence ratio of ρ-state). Due to the condensation, the large positive soft-repulsive potential near the hard sphere contact increases the excess internal energy $u(\rho, \beta)$. As a result, the conflict among u, $\rho(\partial u/\partial \rho)_T$, and p included in α causes switching of α from positive to negative. As a result, negative thermal expansion occurs and the abundance of S-states increases. In other words, when the ratio of the ρ-state grows with cooling, u increases due to a large positive soft-repulsive potential near the hard sphere contact. As a result, α reverses from positive to negative at 4°C, resulting in negative thermal expansion, and the ratio of the S-state increases. That is, the density anomaly is caused by the change in the sign of α, and the change in the abundance of S-state and ρ-state is a result of the density anomaly.

Similarly, we can apply the thermodynamic mechanism of density anomaly elucidated in this book to the other conventional theories [Fomin (2011); Gribova (2009); Molinero and Moore (2009); Oliveira (2009); Sadr (1998)] and the phenomena they insist can be explained briefly. That is, when the energy of the thermal random motion of the molecule decreases with cooling, the energy of expansion motion decreases, so that the water condenses due to the intermolecular attractive force. As a result, the following phenomena could occur: the proportion of the high density component [Röntgen (1892)] could increase, a denser quartz-like structure [Bernal and Fowler (1933)] could be formed, hydrogen bonding distortion [Pople (1951)] could occur, and isolated water molecules could fall into cavities [Pau (1959); Samoilov (1957)]. As the density increases, the positive soft-repulsive potential near the hard sphere contact increases u. As a result, the sign of α changes from positive to negative and it generates negative thermal expansion. Then, with further cooling, it will not be impossible that less dense and ice-like complexes [Röntgen (1892)] increase, less dense tridymite-like ones [Bernal and Fowler (1933)] increase, and the hydrogen bond bending [Pople (1951)] elongates. It is not impossible to enlarge the cavity [Pau (1959); Samoilov (1957)]. In this way, the phenomena discussed in conventional theories can be explained according to the thermodynamic mechanism elucidated in this book. In other words, the phenomena mentioned above should not be regarded as a cause of density anomaly, but as a phenomenon caused by density anomaly.

It is well known that water molecules tend to form a tetrahedral structure. Some people argue that this tetrahedral structure must be the cause of density anomaly. But, is that really the case? As a counter example, consider a folding umbrella. To close and open the umbrella, you must pull or push the root of the skeleton with your hand. The framework itself does not have the driving force to open and close, and the umbrella cannot be opened and closed without the power of a human hand. In order to elucidate the thermodynamic mechanism that generates density anomaly of water, it is necessary to uncover the mysterious force (named as anomaly force for convenience). The anomaly force acts as an intermolecular repulsive force that expands water above 4°C but it acts as an intermolecular attractive force that condenses water below 4°C when heeted. As in the case of a folding umbrella, it is very difficult to imagine that the tetrahedral structure itself has the ability equivalent to anomaly force.

Some say that hydrogen bonds between water molecules are responsible for density anomaly. However, in a thermal equilibrium state, hydrogen bonds act as attractive forces at any temperature. Therefore, it should work only as an action of agglomerating water to increase density. Below 4°C, it is very unlikely that the nature of the hydrogen bond will reverse with cooling and act as a repulsive force that causes water to expand.

Since lactic acid increases with muscle fatigue, it has long been believed that lactic acid is a substance that causes muscle fatigue. However, recently, it has been elucidated that lactic acid is a substance that recovers fatigue, and it is known that the substance that causes fatigue is active oxygen. Fatigue does not occur because lactic acid increases, but lactic acid is secreted to recover from fatigue. Similarly, just because a phenomenon occurs with density anomaly, we cannot say that the phenomenon produces density anomaly. This is true for all theories that have been advocated so far.

Based on the results of experiments and numerical experiments, it has been reported that liquid–liquid phase transition occurs with density anomaly [Mayer and Stanley (1999); Mishima and Stanley (1998); Poole (1992); Sciortino (2003); Sun (2014)]. From this, some people proposed an idea that the liquid–liquid phase transition generates the density anomaly. However, they merely describe the experimental results phenomenologically, they do not explain why the phase transition occurs and why the phase transition produces negative thermal expansion. Since all the phase transitions that occur in normal materials are accompanied by positive thermal expansion, the difference should be explained why only the liquid–liquid phase transition of water produces negative thermal expansion. Just because liquid–liquid phase transition occurs with density anomaly, it cannot be said that liquid–liquid phase transition induces density anomaly. Even if a liquid–liquid phase transition occurs in real water, the author believes that the phase transition may be an accompanying phenomenon as a result of density anomaly caused by the thermodynamic mechanism elucidated in this book.

Summarizing the discussions so far, the author considers the following ideas regarding the density anomaly of water. That is, when negative thermal expansion occurs due to the thermodynamic mechanism explained in Section 2.2 and Section 3.7, it should be considered that a variety of ice structures depending on potential anisotropy, two-state structures, liquid–liquid phase transitions, clathrate structures, etc. are formed by the negative thermal expansion. Anyone of these phenomena should not be considered as the cause of density anomaly, but as a phenomenon that occurs with negative thermal expansion.

The above conclusion is derived from the calculation results obtained using SCOZA. I will explain SCOZA in detail in the following chapters. This theory cannot be applied to freezing or the solid phases of the system because it is a technique built on disordered fluids. However, it is not impossible to get some insight into the thermodynamic properties of water in these phases from the conclusions that this author has drawn. In addition, I believe that the author's conclusions in this book are not limited to water, but are applicable to other phenomena related to density anomaly observed in liquids such as silicon, gallium, germanium, and bismuth.

In the following chapters, after describing the basic knowledge required to understand SCOZA, the basic equations of SCOZA are derived and the numerical calculation method is explained.

Chapter 5

Ornstein–Zernike Equation

As mentioned in Section 1.1, various thermodynamic quantities related to a fluid can be obtained from a radial distribution function $g(r)$. Obtaining $g(r)$ is the biggest challenge of fluid theory This problem results in solving the following two equations:

$$h(r) = c(r) + \rho \int c(r')h(|\mathbf{r} - \mathbf{r}'|)d\mathbf{r}', \tag{5.1}$$

$$c(r) = h(r) - \ln\{1 + h(r)\} - \beta\phi(r) + e(r). \tag{5.2}$$

Here, $h(r) \equiv g(r) - 1$. $h(r)$ is called the total correlation function, and $c(r)$ is called the direct correlation function. Equation (5.1) defines $c(r)$ and is called the Ornstein–Zernike (OZ) equation. Equation (5.2) is called closure. Equations (5.1) and (5.2) are exact fundamental equations of fluid physics. For the specified interparticle interaction $\phi(r)$, $g(r)$ will be obtained by solving these together, but the last term $e(r)$ in Eq. (5.2) makes this difficult. This term is called the bridge function and consists of the sum of all elementary graphs, and no one has succeeded in writing down this term specifically. Therefore, the reliability of fluid theory depends on how accurately $e(r)$ can be captured. Several approximation formulas have been proposed for closure, but this chapter focuses on the following three closures: hyper-netted chain (HNC) approximation, Percus–Yevick (PY) approximation, and mean-spherical approximation (MSA).

The Physics of Liquid Water
Makoto Yasutomi

ISBN 978-981-4877-25-1 (Hardcover), 978-1-003-05616-4 (eBook)
www.jennystanford.com

5.1 Hyper-netted Chain Approximation

In hyper-netted chain (HNC) approximation, we assume $e(r) = 0$. In this case, Eq. (5.2) is given by the following equation:

$$c(r) = h(r) - \ln\{1 + h(r)\} - \beta\phi(r). \tag{5.3}$$

The equation that combines (5.1) and (5.3) is the HNC approximation.

5.2 Percus–Yevick Approximation

Rewrite Eq. (5.3) to get

$$g(r) = \exp[-\beta\phi(r)]\exp[h(r) - c(r)]. \tag{5.4}$$

Expanding the second factor on the right side with $[h(r) - c(r)]$ and ignoring the second and higher terms, we get

$$\begin{aligned} g(r) &= \exp[-\beta\phi(r)]\,[1 + h(r) - c(r))] \\ &= \exp[-\beta\phi(r)]\,[g(r) - c(r))]. \end{aligned} \tag{5.5}$$

Here, $\tau(r)$ is defined by the following equation:

$$\begin{aligned} \tau(r) &= g(r)\mathrm{e}^{\beta\phi(r)}. \\ \therefore g(r) &= \tau(r)\mathrm{e}^{-\beta\phi(r)}. \end{aligned} \tag{5.6}$$

In this case, Eq. (5.5) is given by the following equation:

$$c(r) = \tau(r)\left[\mathrm{e}^{-\beta\phi(r)} - 1\right]. \tag{5.7}$$

Equations (5.6) and (5.7) are closures in PY approximation. As is apparent from the process of deriving Eq. (5.5), this PY approximation is simpler than the HNC approximation, and is rougher.

5.3 Mean-Spherical Approximation

As an intermolecular interaction, we consider an interaction consisting of a hard sphere interaction and tail potential $\phi(r)$. In the

HNC approximation closure (5.3), if the second term on the right-hand side is expanded with $h(r)$ and the second and higher terms are ignored, the closure can be written as follows:

$$h(r) = -1, \qquad r < 1, \tag{5.8}$$

$$c(r) = -\beta\phi(r), \quad r > 1. \tag{5.9}$$

OZ equation (5.1), closures (5.8) and (5.9) are the basic equations of MSA. Equation (5.8) indicates that particles cannot penetrate inside the hard sphere.

5.4 OZ Equations in Baxter Form

In order to solve the OZ equation analytically under MSA, we introduce the unknown function $Q(r)$ and transform the OZ equation (5.1) into a Baxter type [Baxter (1968, 1970)]. In this case, Eq. (5.1) is separated into the following two Baxter-type equations:

$$2\pi rc(r) = -\frac{d}{dr}Q(r) + \rho\int_0^\infty dt\ Q(t)\frac{d}{dr}Q(r+t), \tag{5.10}$$

$$2\pi rh(r) = -\frac{d}{dr}Q(r) + 2\pi\rho\int_0^\infty dt\ Q(t)(r-t)h(|r-t|). \tag{5.11}$$

5.4.1 Baxter Function $Q(r)$

First, the Fourier transform $\tilde{h}(q)$ of the total correlation function $h(r)$ is given by

$$\begin{aligned}
\tilde{h}(q) &= \int h(r)\exp(i\mathbf{q}\cdot\mathbf{r})d\mathbf{r} \\
&= \int_0^\infty dr \int_0^\pi d\theta \int_0^{2\pi} d\phi\ r^2 \sin\theta\ h(r)\exp(iqr\cos\theta) \\
&= 2\pi\int_0^\infty dr\ r^2 h(r)\int_0^\pi d\theta\ \sin\theta\exp(iqr\cos\theta) \\
&= 2\pi\int_0^\infty dr\ r^2 h(r)\int_{-1}^1 dx\ \exp(iqrx) \quad (\text{where } x = \cos\theta) \\
&= 2\pi\int_0^\infty dr\ r^2 h(r)\left[\frac{\exp(iqr)-\exp(-iqr)}{iqr}\right] \\
&= \frac{4\pi}{q}\int_0^\infty dr\ rh(r)\sin(qr).
\end{aligned} \tag{5.12}$$

Here, we define $H(r)$ by the first expression of the following Eq. (5.13) and its differential coefficient is given by the second expression:

$$H(r) = 2\pi \int_r^\infty dx\, xh(x), \quad \frac{dH(r)}{dr} = -2\pi r h(r). \tag{5.13}$$

The partial integration of Eq. (5.12) gives

$$\begin{aligned}
\widetilde{h}(q) &= \frac{4\pi}{q}\int_0^\infty dr \left[-\frac{1}{2\pi}\frac{dH(r)}{dr}\right]\sin(qr) \\
&= \frac{4\pi}{q}\left\{\left[-\frac{1}{2\pi}H(r)\sin(qr)\right]_0^\infty + \frac{q}{2\pi}\int_0^\infty dr\, H(r)\cos(qr)\right\} \\
&= 2\int_0^\infty dr\, H(r)\cos(qr) = \int_{-\infty}^\infty dr\, \exp(iqr)H(|r|).
\end{aligned} \tag{5.14}$$

In summary, the following equation is obtained:

$$\begin{aligned}
\widetilde{h}(q) &= \int h(r)\exp(i\mathbf{q}\cdot\mathbf{r})d\mathbf{r} = \frac{4\pi}{q}\int_0^\infty dr\, rh(r)\sin(qr) \\
&= \int_{-\infty}^\infty dr\, \exp(iqr)H(|r|).
\end{aligned} \tag{5.15}$$

Similarly, $C(r)$ is defined as follows, and the Fourier transform $\widetilde{c}(q)$ of the direct correlation function $c(r)$ is obtained:

$$C(r) = 2\pi \int_r^\infty dx\, xc(x), \tag{5.16}$$

$$\begin{aligned}
\widetilde{c}(q) &= \int c(r)\exp(i\mathbf{q}\cdot\mathbf{r})d\mathbf{r} = \frac{4\pi}{q}\int_0^\infty dr\, rc(r)\sin(qr) \\
&= \int_{-\infty}^\infty dr\, \exp(iqr)C(|r|).
\end{aligned} \tag{5.17}$$

Here, it is assumed that $\widetilde{h}(q)$ and $\widetilde{c}(q)$ depend only on the size of $\mathbf{q}$ and not on the direction.

Multiplying $\exp(i\mathbf{q}\cdot\mathbf{r})$ on both sides of Eq. (5.1) and integrating over the entire region of $\mathbf{r}$, the following formula is obtained:

$$\begin{aligned}
\int h(r)\exp(i\mathbf{q}\cdot\mathbf{r})d\mathbf{r} = &\int c(r)\exp(i\mathbf{q}\cdot\mathbf{r})d\mathbf{r} \\
&+\rho\int d\mathbf{r}'\, c(r')\exp(i\mathbf{q}\cdot\mathbf{r}') \\
&\times\int h(|\mathbf{r}-\mathbf{r}'|)\exp[i\mathbf{q}\cdot(\mathbf{r}-\mathbf{r}')]d\mathbf{r}.
\end{aligned} \tag{5.18}$$

Assume that the fluid is uniform and isotropic. If we set $\mathbf{r}'' = \mathbf{r} - \mathbf{r}'$ for the specified $\mathbf{r}'$, the following expression is obtained:

$$\int h(|\mathbf{r} - \mathbf{r}'|) \exp[i\mathbf{q} \cdot (\mathbf{r} - \mathbf{r}')]d\mathbf{r} = \int h(r'') \exp[i\mathbf{q} \cdot \mathbf{r}'']d\mathbf{r}''$$
$$= \widetilde{h}(q). \tag{5.19}$$

Therefore, Eq. (5.18) is given by the following equation:

$$\widetilde{h}(q) = \widetilde{c}(q) + \rho\widetilde{c}(q)\widetilde{h}(q). \tag{5.20}$$

From Eq. (5.20) we get

$$\widetilde{h}(q) = \frac{\widetilde{c}(q)}{1 - \rho\widetilde{c}(q)}, \quad \widetilde{c}(q) = \frac{\widetilde{h}(q)}{1 + \rho\widetilde{h}(q)}. \tag{5.21}$$

In the above two formulas, multiply each left side and each right side, we get the following:

$$\widetilde{h}(q)\widetilde{c}(q) = \frac{\widetilde{h}(q)\widetilde{c}(q)}{[1 - \rho\widetilde{c}(q)][1 + \rho\widetilde{h}(q)]}. \tag{5.22}$$

Therefore, we get

$$[1 - \rho\widetilde{c}(q)][1 + \rho\widetilde{h}(q)] = 1. \tag{5.23}$$

Using the Wiener–Hopf technique [Baxter (1968, 1970); Nobel (1958); Paley and Wiener (1934)], factorize the first factor on the left side of Eq. (5.23) as follows:

$$1 - \rho\widetilde{c}(q) = Q^{\dagger}(q)Q^{\dagger}(-q), \tag{5.24}$$

where,

$$Q^{\dagger}(q) = 1 - \rho \int_0^{\infty} dr\ Q(r) \exp(iqr). \tag{5.25}$$

Substituting Eq. (5.24) into Eq. (5.23), we get

$$Q^{\dagger}(q)Q^{\dagger}(-q)[1 + \rho\widetilde{h}(q)] = 1. \tag{5.26}$$

Equation (5.26) is the Fourier transformed factorized OZ equation.

Once the unknown function $Q(r)$ is determined, $Q^{\dagger}(q)$ and $\widetilde{c}(q)$ are obtained from Eqs. (5.25) and (5.24). Further, $\widetilde{h}(q)$ is determined from Eq. (5.21). $C(r)$ and $h(r)$ can be obtained by inverse Fourier transform of $\widetilde{c}(q)$ and $\widetilde{h}(q)$.

5.4.2 Baxter's First Equation

The following expression is obtained from Eqs. (5.25), (5.24), and (5.17):

$$
\begin{aligned}
1-\rho\int_{-\infty}^{\infty} dr\,\exp(iqr)C(|r|) &= \left[1-\rho\int_0^{\infty} dr\;Q(r)\exp(iqr)\right]\\
&\times\left[1-\rho\int_0^{\infty} dr\;Q(r)\exp(-iqr)\right]\\
&=1-\rho\int_0^{\infty} dr\;Q(r)\exp(iqr)-\rho\int_0^{\infty} dr\;Q(r)\exp(-iqr)\\
&+\rho^2\int_0^{\infty} dr\int_0^{\infty} dr'\;Q(r)Q(r')\exp[iq(r-r')]. \qquad (5.27)
\end{aligned}
$$

Organize this to get following equation:

$$
\begin{aligned}
\int_{-\infty}^{\infty} dr'\,\exp(iqr')C(|r'|) &= \int_0^{\infty} dr'\;Q(r')\exp(iqr')\\
&+\int_0^{\infty} dr'\;Q(r')\exp(-iqr')\\
&-\rho\int_0^{\infty} dr'\int_0^{\infty} dr''\;Q(r')Q(r'')\exp[iq(r'-r'')], \qquad (5.28)
\end{aligned}
$$

where,

$$
\frac{1}{2\pi}\int_{-\infty}^{\infty} dq\,\exp(iqr)=\delta(r)=\delta(-r). \qquad (5.29)
$$

So, multiply Eq. (5.28) by $\exp(-iqr)/2\pi$ and integrate it with q, the left side, the first term R_1, second term R_2, and third term R_3 of the right side are given by the following equations, respectively,

$$
\begin{aligned}
\text{left side} &= \frac{1}{2\pi}\int_{-\infty}^{\infty} dq\int_{-\infty}^{\infty} dr'\,\exp[iq(r'-r)]C(|r'|)\\
&=\int_{-\infty}^{\infty} dr'\delta(r'-r)C(|r'|)=C(|r|). \qquad (5.30)
\end{aligned}
$$

$$
R_1=\frac{1}{2\pi}\int_{-\infty}^{\infty} dq\int_0^{\infty} dr'\,\exp[iq(r'-r)]Q(r')=Q(r), \qquad (5.31)
$$

$$
R_2=\frac{1}{2\pi}\int_{-\infty}^{\infty} dq\int_0^{\infty} dr'\,\exp[-iq(r'+r)]Q(r')=0, \qquad (5.32)
$$

$$R_3 = -\rho \frac{1}{2\pi} \int_{-\infty}^{\infty} dq \int_{0}^{\infty} dr' \int_{0}^{\infty} dr'' \exp[-iq(r + r'' - r')] \times Q(r')Q(r'') = -\rho \int_{0}^{\infty} dr'' Q(r + r'')Q(r''). \tag{5.33}$$

Therefore, from Eq. (5.28) we get

$$C(|r|) = Q(r) - \rho \int_{0}^{\infty} dr' Q(r + r')Q(r'). \tag{5.34}$$

Differentiate both sides of this equation by r to get

$$2\pi r c(r) = -\frac{dQ(r)}{dr} + \rho \int_{0}^{\infty} dr' \frac{dQ(r + r')}{dr} Q(r'). \tag{5.35}$$

This is the first Baxter equation (5.10) to be found.

5.4.3 Baxter's Second Equation

As shown in Fig. 5.1, consider a single closed curve $C = \Gamma + C_1$ which is composed of two parts of Γ and C_1 on the complex plane $z = x + iy$ ($i = \sqrt{-1}$, x and y: two real numbers). The straight line Γ is $y = 0$, $-R \leq x \leq R$ ($R \to \infty$), and C_1 is a semicircle of radius R centered on the coordinate origin O on the $y \geq 0$ side.

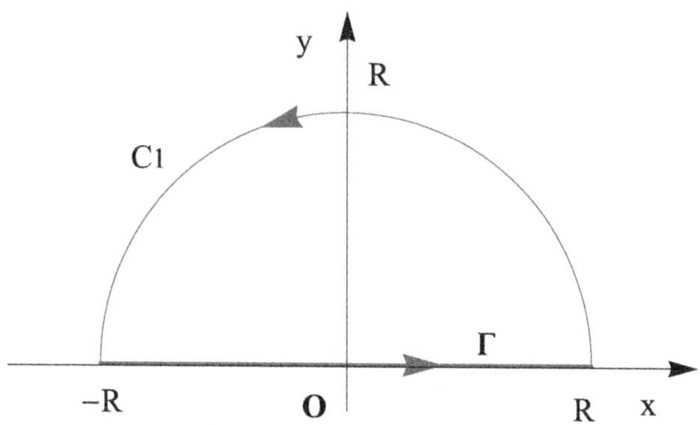

Figure 5.1 Single closed curve C. $C = \Gamma + C_1$ (C has a positive orientation). Straight line Γ: $y = 0$, $-R \leq x \leq R$ ($R \to \infty$), C_1: A semicircle with a radius R centered on the coordinate origin O and on the $y \geq 0$ side.

Assuming $Q^\dagger(z)^{-1}$ is regular in the domain of $y \geq 0$,

$$\frac{1}{2\pi}\int_C dz\ \exp(izr)Q^\dagger(z)^{-1} = \frac{1}{2\pi}\int_{-\infty}^{\infty} dx\ \exp(ixr)Q^\dagger(x)^{-1}$$

$$+\lim_{R\to\infty}\frac{iR}{2\pi}\int_0^{\pi} d\theta\ \exp\left[i(\theta + Rr\cos\theta)\right]\exp\left(-Rr\sin\theta\right) Q^\dagger(-z)^{-1}$$

$$= \frac{1}{2\pi}\int_{-\infty}^{\infty} dx\ \exp(ixr)Q^\dagger(x)^{-1}$$

$$= \frac{1}{2\pi}\int_{-\infty}^{\infty} dx\ \exp(-ixr)Q^\dagger(-x)^{-1} = 0. \tag{5.36}$$

Here, we used $\lim_{R\to\infty} R\exp\left(-Rr\sin\theta\right) = 0$ for $0 < \theta < \pi$. From Eq. (5.26), we get

$$Q^\dagger(q)\left[1 + \rho\tilde{h}(q)\right] = Q^\dagger(-q)^{-1}. \tag{5.37}$$

Multiply $\exp(-iqr)/2\pi$ on both sides of the above formula and integrate q in the region of $-\infty < q < \infty$, from Eq. (5.36) we get

$$\frac{1}{2\pi}\int_{-\infty}^{\infty} dq\ Q^\dagger(q)\left[1 + \rho\tilde{h}(q)\right]\exp(-iqr)$$

$$= \frac{1}{2\pi}\int_{-\infty}^{\infty} dq\ Q^\dagger(q)\exp(-iqr)$$

$$+\frac{\rho}{2\pi}\int_{-\infty}^{\infty} dq\ Q^\dagger(q)\tilde{h}(q)\exp(-iqr) \tag{5.38}$$

$$= \frac{1}{2\pi}\int_{-\infty}^{\infty} dq\ Q^\dagger(-q)^{-1}\exp(-iqr) = 0. \tag{5.39}$$

Substituting Eq. (5.25) for the first term of Eq. (5.38), we get

$$R_{\text{dum}} = \frac{1}{2\pi}\int_{-\infty}^{\infty} dq\ \exp(-iqr)\left[1 - \rho\int_0^{\infty} dr'\ Q(r')\exp(iqr')\right]$$

$$= \delta(r) - \rho\int_0^{\infty} dr'\ Q(r') \times \frac{1}{2\pi}\int_{-\infty}^{\infty} dq\ \exp[iq(r' - r)]$$

$$= -\rho\int_0^{\infty} dr'\ Q(r')\delta(r' - r) = -\rho Q(r) \qquad (r > 0). \tag{5.40}$$

Substitute Eqs. (5.25) and (5.15) for the second term of Eq. (5.38), we get

$$
\begin{aligned}
S_{\text{dum}} &= \frac{\rho}{2\pi}\int_{-\infty}^{\infty} dq \left[1-\rho\int_{0}^{\infty} dr''\; Q(r'')\exp(iqr'')\right] \\
&\quad\times\int_{-\infty}^{\infty} dr' \exp(iqr')H(|r'|)\exp(-iqr) \\
&= \frac{\rho}{2\pi}\int_{-\infty}^{\infty} dq \int_{-\infty}^{\infty} dr' \exp[iq(r'-r)]H(|r'|) \\
&\quad-\frac{\rho^2}{2\pi}\int_{-\infty}^{\infty} dq \int_{0}^{\infty} dr''\; Q(r'')\exp[iq(r''+r'-r)] \\
&\quad\times\int_{-\infty}^{\infty} dr' H(|r'|) \\
&= \rho\int_{-\infty}^{\infty} dr'\delta(r'-r)H(|r'|) \\
&\quad-\rho^2\int_{0}^{\infty} dr''\; Q(r'')\delta(r''+r'-r)\int_{-\infty}^{\infty} dr' H(|r'|) \\
&= \rho H(|r|)-\rho^2\int_{0}^{\infty} dr'' H(|r-r''|)Q(r''). \qquad (5.41)
\end{aligned}
$$

So Eq. (5.38) is written as

$$
-Q(r)+H(|r|)-\rho\int_{0}^{\infty} dr' H(|r-r'|)Q(r')=0. \qquad (5.42)
$$

Differentiate both sides of this equation by r and use Eq. (5.13), we get

$$
2\pi r h(r) = -\frac{dQ(r)}{dr}+2\pi\rho\int_{0}^{\infty} dr'(r-r')h(|r-r'|)Q(r'). \qquad (5.43)
$$

This is the Baxter's second equation (5.11).

5.5 Analytical Solution of the Percus–Yevick Equation for Hard Sphere Fluids

The exact analytical solution of the Percus–Yevick (PY) equation for a hard sphere fluid was first determined independently by [Thiele (1963); Wertheim (1963, 1964)]. Below, the exact solution is derived by following the method of [Baxter (1968)].

5.5.1 Direct Correlation Function

Consider a fluid composed of a hard sphere with a diameter of σ, and its diameter as a unit of length (σ_{u}). Then the interaction between hard spheres is given by

$$\phi(r) = \begin{cases} \infty, & r < 1 \quad (\therefore\ \mathrm{e}^{-\beta\phi(r)} = 0), \\ 0, & r \geq 1 \quad (\therefore\ \mathrm{e}^{-\beta\phi(r)} = 1). \end{cases} \tag{5.44}$$

In this case, PY closure is given by the following equation from Eqs. (5.6) and (5.7):

$$h(r) = \tau(r)\mathrm{e}^{-\beta\phi(r)} - 1 = -1 \quad (r < 1), \tag{5.45}$$

$$c(r) = \tau(r)\left[\mathrm{e}^{-\beta\phi(r)} - 1\right] = 0 \quad (r > 1).$$

Substituting these into Baxter's second equation, we get

$$\begin{aligned} r &= \frac{dQ(r)}{dr} + 2\pi\rho \int_0^1 dt\,(r-t)Q(t) \\ &= \frac{dQ(r)}{dr} + 2\pi\rho \left[r\int_0^1 dt\, Q(t) - \int_0^1 dt\, tQ(t) \right], \end{aligned} \tag{5.46}$$

where, $Q(t) = 0\ (t > 1)$. Now, we set

$$c_a = \int_0^1 dt\, Q(t), \qquad c_b = \int_0^1 dt\, tQ(t), \tag{5.47}$$

and Eq. (5.46) is given by

$$\begin{aligned} r &= \frac{dQ(r)}{dr} + 2\pi\rho(c_a r - c_b) \\ \therefore\ \frac{dQ(r)}{dr} &= (1 - 2\pi\rho c_a)r + 2\pi\rho c_b. \end{aligned} \tag{5.48}$$

Integrate Eq. (5.48) to get

$$Q(r) = \frac{1}{2}(1 - 2\pi\rho c_a)r^2 + 2\pi\rho c_b r + c_c. \tag{5.49}$$

By the way, because $Q(1)=0$,

$$c_c = 6\eta(c_a - 2c_b) - \frac{1}{2}, \qquad \left(\eta = \frac{\pi\rho}{6}\right). \tag{5.50}$$

From Eqs. (5.47), (5.49), and (5.50), we get

$$Q(r) = -\frac{1}{2(1-\eta)} - \frac{3\eta}{2(1-\eta)^2}r + \frac{1+2\eta}{2(1-\eta)^2}r^2. \tag{5.51}$$

Substituting $Q(r)$ into Baxter's first equation, we get

$$rc(r) = -\frac{dQ(r)}{dr} + 2\pi\rho\int_r^1 dt\,\frac{dQ(t)}{dt}Q(t-r). \tag{5.52}$$

After integrating the above equation and rearranging it, the following equation is obtained for $(0 < r \le 1)$:

$$\begin{aligned} c(r) = -\tau(r) &= -\frac{(1+2\eta)^2}{(1-\eta)^4} + \frac{6\eta(1+\eta/2)^2}{(1-\eta)^4}r \\ &-\frac{\eta(1+2\eta)^2}{2(1-\eta)^4}r^3. \end{aligned} \tag{5.53}$$

5.5.2 Equation of State Derived from Virial Theorem

From Eqs. (1.9) and (5.6), we get

$$\frac{pV}{\mathcal{N}kT} - 1 - \frac{2\pi\rho}{3kT}\int_0^\infty \frac{d\phi(r)}{dr}\tau(r)e^{-\beta\phi(r)}r^2dr. \tag{5.54}$$

By the way,

$$\frac{d\phi(r)}{dr}e^{-\beta\phi(r)} = -\frac{1}{\beta}\frac{d}{dr}e^{-\beta\phi(r)} = -kT\,\delta(r-1). \tag{5.55}$$

Here, $\delta(r-1)$ is a Dirac delta function. Therefore, the following expression is obtained from Eqs. (5.53), (5.54), and (5.55):

$$\begin{aligned} \frac{pV}{\mathcal{N}kT} &= 1 + \frac{2\pi\rho}{3}\int_0^\infty \tau(r)r^2\delta(r-1)dr = 1 + \frac{2\pi\rho}{3}\tau(1) \\ &= \frac{1+2\eta+3\eta^2}{(1-\eta)^2} = \frac{1+\eta+\eta^2-3\eta^3}{(1-\eta)^3}. \end{aligned} \tag{5.56}$$

5.5.3 Equation of State Derived from Compressibility

Rewrite the compressibility equation (1.7) using Eqs. (5.45) and (5.44) to get

$$\begin{aligned} kT\left(\frac{\partial\rho}{\partial p}\right)_T &= 1 + 4\pi\rho\int_0^\infty [g(r)-1]\,r^2dr \\ &= 1 + 4\pi\rho\int_0^\infty \left[\tau(r)e^{-\beta\phi(r)} - 1\right]r^2dr \\ &= 1 + 4\pi\rho\left[\int_0^1 \left[\tau(r)e^{-\beta\phi(r)} - 1\right]r^2dr \right.\\ &\quad \left. + \int_1^\infty \left[\tau(r)e^{-\beta\phi(r)} - 1\right]r^2dr\right] \end{aligned}$$

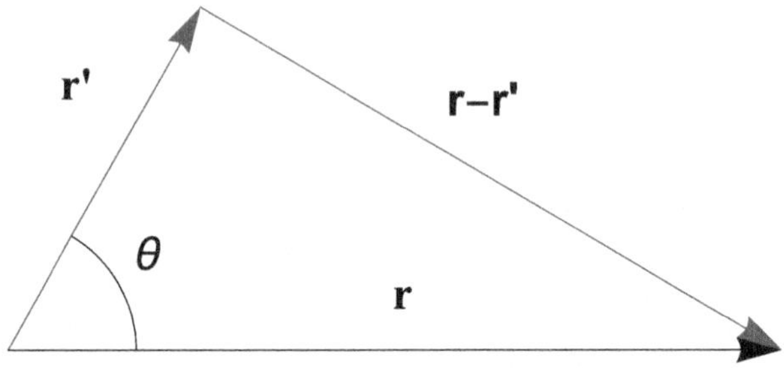

Figure 5.2

$$= 1 + 4\pi\rho\left[-\int_0^1 r^2 dr + \int_1^\infty [\tau(r) - 1] r^2 dr \right]$$
$$= 1 - \frac{4\pi\rho}{3} + 4\pi\rho \int_1^\infty [\tau(r) - 1] r^2 dr. \tag{5.57}$$

As for $\tau(r)$, it is described for $(0 < r \leq 1)$ in Eq. (5.53), but it is still not available for $(r > 1)$. So, find the integral of the last term in the above equation.

From PY closure and OZ equation (5.1), we get (see Fig. 5.2)

$$h(r) - c(r) = \tau(r) - 1 = \rho \int d\mathbf{r}' c(r') h(|\mathbf{r} - \mathbf{r}'|)$$
$$= \rho \int d\mathbf{r}' \tau(r') \left[e^{-\beta\phi(r')} - 1 \right] \left[-1 + \tau(|\mathbf{r} - \mathbf{r}'|) e^{-\beta\phi(|\mathbf{r}-\mathbf{r}'|)} \right]$$
$$= -\rho \int d\mathbf{r}' \tau(r') \left[e^{-\beta\phi(r')} - 1 \right]$$
$$+ \rho \int d\mathbf{r}' \tau(r') \left[e^{-\beta\phi(r')} - 1 \right] \tau(|\mathbf{r} - \mathbf{r}'|) e^{-\beta\phi(|\mathbf{r}-\mathbf{r}'|)}$$
$$= -4\pi\rho \int_0^\infty dr' r'^2 \tau(r') \left[e^{-\beta\phi(r')} - 1 \right]$$
$$+ 2\pi\rho \int_0^\pi \sin\theta d\theta \int_0^\infty dr' \, r'^2 \tau(r') \left[e^{-\beta\phi(r')} - 1 \right]$$
$$\times \tau(|\mathbf{r} - \mathbf{r}'|) e^{-\beta\phi(|\mathbf{r}-\mathbf{r}'|)}. \tag{5.58}$$

Here, we set

$$\hat{A} = \int_0^1 dx \, x^2 \tau(x) = \frac{(4 - \eta)(2 + \eta^2)}{24(1 - \eta)^4}. \tag{5.59}$$

The integration of the first term on the right-hand side of Eq. (5.58) can be done using Eq. (5.44), and we get

$$T_{\text{dum}} = -4\pi\rho \int_0^\infty dr' r'^2 \tau(r') \left[\mathrm{e}^{-\beta\phi(r')} - 1 \right] = 4\pi\rho\hat{A}. \tag{5.60}$$

In the integration of the second term on the right-hand side of Eq. (5.58), convert the variable from θ to t to get the relations

$$t^2 = s^2 + r^2 - 2rs\cos\theta \quad \rightarrow \quad tdt = rs\sin\theta d\theta$$

where, $|\mathbf{r}'| = s$, $|\mathbf{r} - \mathbf{r}'| = t$. And we get the following:

$$\begin{aligned} \mathrm{R}_2 &= \frac{2\pi\rho}{r} \int_0^\infty ds\, s\tau(s) \left[\mathrm{e}^{-\beta\phi(s)} - 1 \right] \int_{|r-s|(>1)}^{r+s(>1)} dt\, t\tau(t) \mathrm{e}^{-\beta\phi(t)} \\ &= -\frac{2\pi\rho}{r} \int_0^1 ds\, s\tau(s) \int_{|r-s|(>1)}^{r+s(>1)} dt\, t\tau(t). \end{aligned} \tag{5.61}$$

From Eqs. (5.58), (5.60), and (5.61), we get

$$\tau(r) - 1 = 4\pi\rho\hat{A} - \frac{2\pi\rho}{r} \int_0^1 ds\, s\tau(s) \int_{|r-s|(>1)}^{r+s(>1)} dt\, t\tau(t). \tag{5.62}$$

By the way,

$$\begin{aligned} \int_1^\infty dr\, r^2 \left[\tau(r) - 1 \right] &= \int_0^\infty dr\, r^2 \left[\tau(r) - 1 \right] \\ &\quad - \int_0^1 dr\, r^2 \left[\tau(r) - 1 \right]. \end{aligned} \tag{5.63}$$

Here, we set

$$\hat{B} = \int_1^\infty dr\, r^2 \left[\tau(r) - 1 \right], \qquad \hat{C} = \int_0^\infty dr\, r^2 \left[\tau(r) - 1 \right]. \tag{5.64}$$

Now, Eq. (5.63) is given by

$$\hat{B} = \hat{C} - \hat{A} + \frac{1}{3}. \tag{5.65}$$

Using Eq. (5.62), $\hat{C}$ is written as

$$\begin{aligned} \hat{C} &= 4\pi\rho\hat{A} \int_0^\infty dr\, r^2 \\ &\quad -2\pi\rho \int_0^1 ds\, s\tau(s) \int_0^\infty dr\, r \int_{|r-s|(>1)}^{r+s(>1)} dt\, t\tau(t). \end{aligned} \tag{5.66}$$

The first term F_1 on the right-hand side of Eq. (5.66) is

$$F_1 = 4\pi\rho\hat{A}\left(\int_0^1 dr\, r^2 + \int_1^\infty dr\, r^2\right)$$
$$= 4\pi\rho\hat{A}\left(\frac{1}{3} + \int_1^\infty dr\, r^2\right). \quad (5.67)$$

As for the second term F_2 on the right-hand side of Eq. (5.66), dividing the integral over r into $0 \leq r \leq 1$ and $1 \leq r \leq \infty$, we get

$$F_2 = -2\pi\rho \int_0^1 ds\, s\tau(s) \int_0^\infty dr\, r \int_{|r-s|(>1)}^{r+s(>1)} dt\, t\tau(t)$$
$$= -2\pi\rho \int_0^1 ds\, s\tau(s) \int_0^1 dr\, r \int_{|r-s|(>1)}^{r+s(>1)} dt\, t\tau(t)$$
$$-2\pi\rho \int_0^1 ds\, s\tau(s) \int_1^\infty dr\, r \int_{|r-s|(>1)}^{r+s(>1)} dt\, t\tau(t). \quad (5.68)$$

As for the first term F_{21} on the right-hand side of Eq. (5.68), dividing the integral over r into $0 \leq r \leq s$ and $s \leq r \leq 1$, we get

$$-\frac{F_{21}}{2\pi\rho} = \int_0^1 ds\, s\tau(s) \int_0^1 dr\, r \int_{|r-s|(>1)}^{r+s(>1)} dt\, t\tau(t)$$
$$= \int_0^1 ds\, s\tau(s) \int_0^s dr\, r \int_{s-r(>1)}^{r+s(>1)} dt\, t\tau(t)$$
$$+ \int_0^1 ds\, s\tau(s) \int_s^1 dr\, r \int_{r-s(>1)}^{r+s(>1)} dt\, t\tau(t). \quad (5.69)$$

In the first term on the right-hand side of Eq. (5.69), the following equation is obtained by switching the integration order of r and t (see Section A.1):

$$-\frac{F_{211}}{2\pi\rho} = \int_0^1 ds\, s\tau(s) \int_0^s dr\, r \int_{s-r(>1)}^{r+s(>1)} dt\, t\tau(t)$$
$$= \int_{1/2}^1 ds\, s\tau(s) \int_1^{2s} dt\, t\tau(t) \int_{t-s}^s dr\, r.$$

Similarly, in the second term on the right-hand side of Eq. (5.69), the following equation is obtained (see Section A.2):

$$-\frac{F_{212}}{2\pi\rho} = \int_0^1 ds\, s\tau(s) \int_s^1 dr\, r \int_{r-s(>1)}^{r+s(>1)} dt\, t\tau(t)$$

$$= \int_0^{1/2} ds\, s\tau(s) \int_1^{1+s} dt\, t\tau(t) \int_{t-s}^1 dr\, r$$

$$+ \int_{1/2}^1 ds\, s\tau(s) \left[\int_1^{2s} dt\, t\tau(t) \int_s^1 dr\, r + \int_{2s}^{1+s} dt\, t\tau(t) \int_{t-s}^1 dr\, r \right].$$

From $F_{21} = F_{211} + F_{212}$, we get

$$\begin{aligned}
-\frac{F_{21}}{2\pi\rho} &= -\frac{F_{211} + F_{212}}{2\pi\rho} \\
&= \int_{1/2}^1 ds\, s\tau(s) \int_1^{2s} dt\, t\tau(t) \left[\int_s^1 dr + \int_{t-s}^s dr\, r \right] \\
&\quad + \int_0^{1/2} ds\, s\tau(s) \int_1^{1+s} dt\, t\tau(t) \int_{t-s}^1 dr\, r \\
&\quad + \int_{1/2}^1 ds\, s\tau(s) \int_{2s}^{1+s} dt\, t\tau(t) \int_{t-s}^1 dr\, r \\
&= \int_{1/2}^1 ds\, s\tau(s) \int_1^{2s} dt\, t\tau(t) \int_{t-s}^1 dr\, r \\
&\quad + \int_0^{1/2} ds\, s\tau(s) \int_1^{1+s} dt\, t\tau(t) \int_{t-s}^1 dr\, r \\
&\quad + \int_{1/2}^1 ds\, s\tau(s) \int_{2s}^{1+s} dt\, t\tau(t) \int_{t-s}^1 dr\, r \\
&= \int_{1/2}^1 ds\, s\tau(s) \int_1^{1+s} dt\, t\tau(t) \int_{t-s}^1 dr\, r \\
&\quad + \int_0^{1/2} ds\, s\tau(s) \int_1^{1+s} dt\, t\tau(t) \int_{t-s}^1 dr\, r \\
&= \int_0^1 ds\, s\tau(s) \int_1^{1+s} dt\, t\tau(t) \int_{t-s}^1 dr\, r.
\end{aligned} \tag{5.70}$$

The second term F_{22} on the right-hand side of Eq. (5.68) can be obtained by switching the integration order of r and t (see

Section A.3).

$$\begin{aligned}
-\frac{F_{22}}{2\pi\rho} &= \int_0^1 ds\, s\tau(s) \int_1^\infty dr\, r \int_{r-s(>1)}^{r+s(>1)} dt\, t\tau(t) \\
&= \int_0^1 ds\, s\tau(s) \left[\int_1^{1+s} dt\, t\tau(t) \int_1^{t+s} dr\, r \right. \\
&\quad \left. + \int_{1+s}^\infty dt\, t\tau(t) \int_{t-s}^{t+s} dr\, r \right].
\end{aligned} \tag{5.71}$$

So Eq. (5.66) is

$$\begin{aligned}
\hat{C} &= 4\pi\rho\hat{A} \int_0^\infty dr\, r^2 - 2\pi\rho(F_{21} + F_{22}) \\
&= 4\pi\rho\hat{A} \left[\frac{1}{3} + \int_1^\infty dr\, r^2 \right] \\
&\quad -2\pi\rho \left[\int_0^1 ds\, s\tau(s) \int_1^{1+s} dt\, t\tau(t) \int_{t-s}^1 dr\, r \right. \\
&\quad + \int_0^1 ds\, s\tau(s) \int_1^{1+s} dt\, t\tau(t) \int_1^{t+s} dr\, r \\
&\quad \left. + \int_0^1 ds\, s\tau(s) \int_{1+s}^\infty dt\, t\tau(t) \int_{t-s}^{t+s} dr\, r \right] \\
&= 4\pi\rho\hat{A} \left[\frac{1}{3} + \int_1^\infty dr\, r^2 \right] \\
&\quad -2\pi\rho \left[\int_0^1 ds\, s\tau(s) \int_1^{1+s} dt\, t\tau(t) \int_{t-s}^{t+s} dr\, r \right. \\
&\quad \left. + \int_0^1 ds\, s\tau(s) \int_{1+s}^\infty dt\, t\tau(t) \int_{t-s}^{t+s} dr\, r \right] \\
&= 4\pi\rho\hat{A} \left[\frac{1}{3} + \int_1^\infty dr\, r^2 \right] \\
&\quad -2\pi\rho \int_0^1 ds\, s\tau(s) \int_1^\infty dt\, t\tau(t) \int_{t-s}^{t+s} dr\, r.
\end{aligned}$$

Performing the third integral of the second term on the rightmost side of the above equation and using Eqs. (5.59) and (5.64), we get

$$\hat{C} = 4\pi\rho\hat{A} \left[\frac{1}{3} + \int_1^\infty dr\, r^2 \right] - 4\pi\rho \int_0^1 ds\, s^2\tau(s) \int_1^\infty dt\, t^2\tau(t)$$

and

$$\hat{C} = \frac{4\pi\rho}{3}\hat{A} + 4\pi\rho\hat{A}\int_1^\infty dr\, r^2 - 4\pi\rho\hat{A}\int_1^\infty dr\, r^2\tau(r)$$
$$= 4\pi\rho\hat{A}\left(\frac{1}{3} - \hat{B}\right). \tag{5.72}$$

From Eqs. (5.72) and (5.65),

$$\hat{B} = \frac{1}{1 + 24\eta\hat{A}}\left[(8\eta - 1)\hat{A} + \frac{1}{3}\right]. \tag{5.73}$$

From Eqs. (5.57), (5.64), and (5.73)

$$kT\left(\frac{\partial\rho}{\partial p}\right)_T = \frac{1}{1 + 24\eta\hat{A}}. \tag{5.74}$$

Divide the reciprocal of Eq. (5.74) by ρ and then integrate over ρ to get

$$\frac{pV}{\mathcal{N}kT} = \frac{1}{\rho kT}\int_0^p d\rho\left(\frac{\partial p}{\partial\rho}\right)_T$$
$$= \frac{1}{\eta}\int_0^\eta d\eta\,(1 + 24\eta\hat{A})$$
$$= \frac{1 + \eta + \eta^2}{(1 - \eta)^3}. \tag{5.75}$$

This is the equation of state derived from compressibility for hard sphere fluid under PY approximation.

5.5.4 Carnahan–Starling's Empirical Formula

Carnahan and Starling obtained the following equation to reproduce the equation of state of a hard sphere fluid obtained from numerical experiments with high accuracy:

$$\frac{pV}{\mathcal{N}kT} = \frac{1 + \eta + \eta^2 - \eta^3}{\Delta^3}, \quad \text{where,} \quad \Delta = 1 - \eta. \tag{5.76}$$

The deviation from the numerical experiment of this formula is less than 0.3%. We show Carnahan–Starling equation (5.76), equations of state (5.56) and (5.75) in Fig. 5.3. If the fluid theory is exact, the equations of state should all agree regardless of the derived path. However, when comparing Eqs. (5.56) and (5.75), they

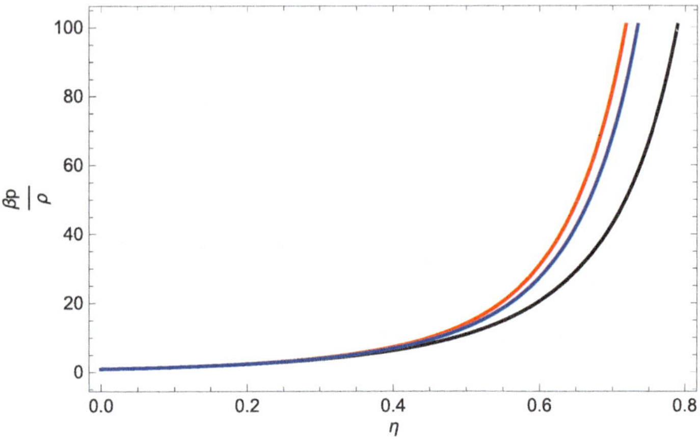

Figure 5.3 Equation of state of hard sphere fluid. Red line: Equation of state obtained from compression ratio. Black line: Equation of state obtained from Virial theorem. Blue line: Carnahan–Starling empirical formula [Carnahan and Starling (1969)].

agree on the low-density side, but the deviation increases toward the high-density side. This is a contradiction caused by a term ignored when obtaining a PY closure, as explained in Section 5.2.

As explained in Chapter 1, there are three paths of compressibility equation (1.7), energy equation (1.8), and virial theorem to get the state equation. However, at present, there is no simplified model that satisfies all the equations obtained from these three paths. In this chapter, we develop a discussion based on the results obtained by using self-consistent Ornstein–Zernike approximation (SCOZA) [Betancourt-Cárdenas (2008); Pini (1998a,b); Schöll-Paschinger (2004, 2005); Yasutomi (2010, 2014, 2015, 2016)]. SCOZA is a theory built to satisfy the compressibility equation and energy equation consistently. It uses the analytical solution obtained by solving the Baxter OZ equations (5.10) and (5.11) under MSA.

5.6 Self-Consistent OZ Approximation

The Baxter-form OZ equations (5.10) and (5.11) can be solved analytically under the MSA for the following seven different types of interparticle interactions:

(1) Neutral hard-sphere interaction [Henderson (1976, 1980); Plischke and Henderson (1986); Thompson (1980); Waisman (1976)].
(2) Sticky hard-sphere interaction [Barboy and Tenne (1979); Baxter (1968); Ginoza and Yasutomi (1996); Perram and Smith (1975)].
(3) Yukawa interaction [Blum and Hoye (1978); Ginoza (1985, 1986); Hoye and Blum (1977); Waisman (1973)].
(4) Hard ions and Yukawa interaction [Blum (1980)].
(5) Sticky hard-sphere Yukawa interaction [Yasutomi and Ginoza (1996)].
(6) Sogami-Ise interaction [Yasutomi and Ginoza (2000)].
(7) Screened Coulomb plus power series interaction [Yasutomi (2001)].

However, as described in Chapter 1, the MSA has a rough approximation. The equations of state derived from each of the three different routes are not completely consistent, and the deviation increases with decreasing temperature. The theory proposed to improve such inconvenience is SCOZA. This chapter explains this theory.

5.6.1 SCOZA Closure

Consider the following closure for the Baxter OZ equations (5.10) and (5.11):

$$h(r) = -1, \qquad r < 1, \tag{5.77}$$

$$c(r) = c_{\text{HS}}(r) - R(\rho, \beta)\beta\phi(r), \qquad r \geq 1. \tag{5.78}$$

The total correlation function $h(r)$ for a hard sphere in Eq. (5.77) indicates that particles cannot penetrate $r < 1$. The function $c_{\text{HS}}(r)$ of Eq. (5.78) represents the direct correlation function [Pini (1998)] of a hard sphere fluid and is given by the following equation:

$$c_{\text{HS}}(r) = \mathcal{K}^{(1)}(\rho)\frac{\exp[-z_1(\rho)(r-1)]}{r} \qquad r > 1. \tag{5.79}$$

Here, $\mathcal{K}^{(1)}(\rho)$ and $z_1(\rho)$ are known functions of ρ (see Section 6.2). The quantity $R(\rho, \beta)$ in Eq. (5.78) is a correction factor introduced to improve the contradiction generated by the roughness of the MSA approximation. $R(\rho, \beta)$ is used to match the two equations of state derived from the compressibility equation (1.7) and the energy equation (1.8). It can be obtained by solving the conditional expression (1.24).

When the intermolecular interaction $\phi(r)$ is given by multi-Yukawa terms (3.2) or screened power-series interaction [Yasutomi (2010)], the analytical solution for each intermolecular interaction can be imported into SCOZA. Arbitrary smooth functions can be developed with multi-Yukawa terms or screened power-series, so SCOZA in this case can be applied to interactions of various shapes. However, this book adopts the former closure and rewrites the analytical solution for the mixed component system of [Blum and Hoye (1978)] to a single component system to be incorporated into the present SCOZA.

5.6.2 Laplace Transformation of Baxter-Type OZ Equation

Here, we find analytical solutions of Baxter-format OZ equations (5.10) and (5.11) following the method of [Blum and Hoye (1978)] and using the Laplace transform and its inverse. In this case, since the distribution function $g(r)$ is discontinuous at $r = 1$, careful handling is required in the vicinity thereof. We define $1^+ = \lim_{\epsilon\to 0}(1+|\epsilon|)$, $1^- = \lim_{\epsilon\to 0}(1-|\epsilon|)$, and the following equation holds:

$$g(1^+) > 0, \quad g(1^-) = 0. \tag{5.80}$$

5.6.3 Laplace Transformation of Baxter's Second Equation

First, the Laplace transform of Eq. (5.11) is obtained.

- In the case of $r > 0$.

Substituting $h(r) = g(r) - 1$ into the second term on the right-hand side of Eq. (5.11) gives

$$
\begin{aligned}
&2\pi\rho\int_0^\infty dt\; Q(t)(r-t)h(|r-t|)\\
&= 2\pi\rho\int_0^\infty dt\; Q(t)(r-t)\,[g(|r-t|)-1]\\
&= 2\pi\rho\int_0^\infty dt\; Q(t)(r-t)g(|r-t|) - 2\pi\rho\int_0^\infty dt\; Q(t)(r-t)\\
&= 2\pi\rho\int_0^\infty dt\; Q(t)(r-t)g(|r-t|) + \left(-2\pi\rho\int_0^\infty dt\; Q(t)\right) r\\
&\quad + 2\pi\rho\int_0^\infty dt\; Q(t)t\\
&= 2\pi\rho\int_0^\infty dt\; Q(t)(r-t)g(|r-t|) + (A-2\pi)r + B, \qquad (5.81)
\end{aligned}
$$

where,

$$
A-2\pi \equiv -2\pi\rho\int_0^\infty dt\; Q(t), \quad B \equiv 2\pi\rho\int_0^\infty dt\; Q(t)t. \qquad (5.82)
$$

Substituting Eq. (5.81) into Eq. (5.11), we get the following for $(r>0)$

$$
\begin{aligned}
2\pi r h(r) = &-\frac{d}{dr}Q(r) + 2\pi\rho\int_0^\infty dt\; Q(t)(r-t)g(|r-t|)\\
&+(A-2\pi)r + B. \qquad (5.83)
\end{aligned}
$$

Since $h(r) = g(r) - 1$, the above equation is applied as follows:

$$
2\pi r g(r) = -\frac{d}{dr}Q(r) + 2\pi\rho\, G_{\mathrm{dum}} + Ar + B, \qquad (5.84)
$$

$$
\text{where,}\quad G_{\mathrm{dum}} = \int_0^\infty dt\; Q(t)(r-t)g(|r-t|).
$$

- In the case of $0<r<1$.

From Eq. (5.8), $g(r)=0$ when $r<1$, so the variable domain in the integral of G_{dum} is

$$
(r-1>t \text{ or } t>1+r) \text{ and } t>0 \text{ and } 0<r<1.
$$

Therefore,

$$
(r-1>t \text{ or } t>1+r) \text{ and } t>0 \text{ and } 0<r<1.
$$

And, $t > r + 1$, so

$$G_{\text{dum}} = \int_{1+r}^{\infty} dt\ Q(t)(r-t)g(t-r). \tag{5.85}$$

Here, if $x = t - r$ is set, $1 < x < \infty$. Therefore, above equation gives

$$G_{\text{dum}} = -\int_{1}^{\infty} dx\ Q(x+r)\,xg(x). \tag{5.86}$$

Now, for any positive integer, we set the following

$$\begin{cases} Q(r) \equiv Q^{(0)}(r) + Q^{(1)}(r), \\ Q^{(0)}(r) = 0, \qquad r > 1, \\ Q^{(1)}(r) \equiv \sum_{n=1}^{N} D^{(n)} e^{-z_n r}, \quad r > 0. \end{cases} \tag{5.87}$$

Substituting Eq. (5.87) for Eq. (5.86), we get

$$\begin{aligned} G_{\text{dum}} &= -\int_{1}^{\infty} dx\ Q^{(1)}(x+r)\,xg(x) \\ &= -\sum_{n=1}^{N} D^{(n)} \mathrm{e}^{-z_n r} \int_{1}^{\infty} dx\ \mathrm{e}^{-z_n x}\ xg(x) \\ &= -\sum_{n=1}^{N} D^{(n)} \mathrm{e}^{-z_n r}\ \tilde{g}(z_n). \end{aligned} \tag{5.88}$$

Here,

$$\tilde{g}(z_n) \equiv \int_{1}^{\infty} dx\ \mathrm{e}^{-z_n x}\ xg(x) = \int_{0}^{\infty} dx\ \mathrm{e}^{-z_n x}\ xg(x). \tag{5.89}$$

Assign Eq. (5.88) to Eq. (5.84) to get

$$2\pi r g(r) = -\frac{d}{dr}Q(r) - 2\pi\rho \sum_{n=1}^{N} D^{(n)} \mathrm{e}^{-z_n r}\ \tilde{g}(z_n) + Ar + B, \tag{5.90}$$

where, $(0 < r < 1^{+})$. Here we introduce $C^{(n)}$ that satisfies

$$z_n(C^{(n)} + D^{(n)}) = 2\pi\rho D^{(n)}\ \tilde{g}(z_n). \tag{5.91}$$

Substituting this expression into Eq. (5.90) gives the following expression:

$$2\pi r g(r) = -\frac{d}{dr}Q(r) - \sum_{n=1}^{N} z_n \mathrm{e}^{-z_n r}(C^{(n)} + D^{(n)}) + Ar + B, \tag{5.92}$$

where, $(0 < r < 1^+)$. In $0 < r < 1^-, g(r) = 0$, so from Eqs. (5.92) and (5.87), we get

$$\begin{aligned}\frac{d}{dr}Q^{(0)}(r) &= -\frac{d}{dr}Q^{(1)}(r) - \sum_{n=1}^{N} z_n \mathrm{e}^{-z_n r}(C^{(n)} + D^{(n)}) + Ar + B \\ &= \sum_{n=1}^{N} z_n \mathrm{e}^{-z_n r} D^{(n)} - \sum_{n=1}^{N} z_n \mathrm{e}^{-z_n r}(C^{(n)} + D^{(n)}) + Ar + B \\ &= -\sum_{n=1}^{N} z_n \mathrm{e}^{-z_n r} C^{(n)} + Ar + B. \end{aligned} \tag{5.93}$$

Because $Q^{(0)}(1) = 0$, integration of Eq. (5.93) with respect to r from r (< 1) to 1^- gives the following:

$$\begin{aligned}Q^{(0)}(r) &= \frac{1}{2}A(r^2 - 1) + B(r - 1) \\ &\quad + \sum_{n=1}^{N} C^{(n)}(\mathrm{e}^{-z_n r} - \mathrm{e}^{-z_n}), \quad (0 < r < 1). \end{aligned} \tag{5.94}$$

- In the case of $1 < r$.

If $1 < r$ in Eqs. (5.84) and (5.87), we get

$$\begin{aligned}2\pi r g(r) &= -\frac{d}{dr}Q^{(1)}(r) + 2\pi\rho \int_0^\infty dt\, Q(t)(r - t)g(|r - t|) \\ &\quad + Ar + B, \quad (1 < r). \end{aligned} \tag{5.95}$$

The operator $\int_1^\infty dr\ \mathrm{e}^{-sr}$ is applied to both sides of the above expression to obtain the following:

$$2\pi\tilde{g}(s) = F_{\mathrm{dum}}^{(1)} + F_{\mathrm{dum}}^{(2)} + F_{\mathrm{dum}}^{(3)}, \tag{5.96}$$

where,

$$\begin{aligned}F_{\mathrm{dum}}^{(1)} &= -\int_1^\infty dr\, \mathrm{e}^{-sr}\frac{d}{dr}Q^{(1)}(r) = \sum_{n=1}^{N} z_n D^{(n)} \int_1^\infty dr\, \mathrm{e}^{-(s+z_n)r} \\ &= \sum_{n=1}^{N} z_n D^{(n)} \frac{\mathrm{e}^{-(s+z_n)}}{s + z_n}, \end{aligned} \tag{5.97}$$

$$F_{\mathrm{dum}}^{(3)} = \int_1^\infty dr\, \mathrm{e}^{-sr}(Ar + B) = \frac{\mathrm{e}^{-s}[A + (A + B)s]}{s^2}. \tag{5.98}$$

$$\begin{aligned}
F_{\text{dum}}^{(2)} &= 2\pi\rho \int_0^\infty dt\, Q(t) \int_1^\infty dr\, \mathrm{e}^{-sr}(r-t)g(|r-t|) \\
&= 2\pi\rho \int_0^r dt\, Q(t) \int_1^\infty dr\, \mathrm{e}^{-sr}(r-t)g(|r-t|) \\
&\quad +2\pi\rho \int_r^\infty dt\, Q(t) \int_1^\infty dr\, \mathrm{e}^{-sr}(r-t)g(|r-t|) \\
&= 2\pi\rho \left[F_{\text{dum}}^{(21)} + F_{\text{dum}}^{(22)} \right], \qquad (5.99)
\end{aligned}$$

where,

$$\begin{aligned}
F_{\text{dum}}^{(22)} &= \int_r^\infty dt\, Q(t) \int_1^\infty dr\, \mathrm{e}^{-sr}(r-t)g(|r-t|) \\
&= \int_1^\infty dr\, \mathrm{e}^{-sr} \int_r^\infty dt\, Q^{(1)}(t)(r-t)g(t-r) \\
&= \sum_{n=1}^{N} D^{(n)} \int_1^\infty dr\, \mathrm{e}^{-sr} \int_r^\infty dt\, \mathrm{e}^{-z_n t}(r-t)g(t-r).
\end{aligned}$$

In the above integration, if you set $x = t - r$,

$$\begin{aligned}
F_{\text{dum}}^{(22)} &= -\sum_{n=1}^{N} D^{(n)} \int_1^\infty dr\, \mathrm{e}^{-(z_n+s)r} \int_0^\infty dx\, \mathrm{e}^{-z_n x} x g(x) \\
&= -\sum_{n=1}^{N} D^{(n)} \tilde{g}(z_n) \frac{\mathrm{e}^{-(z_n+s)}}{z_n+s}. \qquad (5.100)
\end{aligned}$$

And, $F_{\text{dum}}^{(21)}$ is given by

$$\begin{aligned}
F_{\text{dum}}^{(21)} &= \int_0^r dt\, Q(t) \int_1^\infty dr\, \mathrm{e}^{-sr}(r-t)g(|r-t|) \\
&= \int_0^r dt\, Q(t) \left[\int_0^\infty dr\, \mathrm{e}^{-sr}(r-t)g(|r-t|) \right. \\
&\quad \left. - \int_0^1 dr\, \mathrm{e}^{-sr}(r-t)g(|r-t|) \right] \\
&= \int_0^\infty dr\, \mathrm{e}^{-sr} \int_0^r dt\, Q(t)(r-t)g(r-t) \\
&\quad - \int_0^1 dr\, \mathrm{e}^{-sr} \int_0^r dt\, Q(t)(r-t)g(r-t) \\
&= \int_0^\infty dr\, \mathrm{e}^{-sr} \int_0^r dt\, Q(t)(r-t)g(r-t). \qquad (5.101)
\end{aligned}$$

This is because $g(r-t) = 0$ in the second row from the bottom in the above expression. In that line, since $r < 1$ and $0 < t < r, r - t < 1$. Therefore, $g(r - t) = 0$ from Eq. (5.77).

From the Laplace transform inversion formula for Eq. (5.89), for real γ, we get

$$(r-t)\,g(r-t) = \frac{1}{2\pi i}\int_{\gamma-i\infty}^{\gamma+i\infty} dp\, \mathrm{e}^{p(r-t)}\,\tilde{g}(p). \tag{5.102}$$

Substitute Eq. (5.102) for the rightmost side of Eq. (5.101), change the integration order, and perform partial integration to obtain the following expression:

$$\begin{aligned} F_{\mathrm{dum}}^{(21)} &= \int_0^\infty dr\, \mathrm{e}^{-sr}\int_0^r dt\, Q(t)\,\frac{1}{2\pi i}\int_{\gamma-i\infty}^{\gamma+i\infty} dp\, \mathrm{e}^{p(r-t)}\,\tilde{g}(p) \\ &= \frac{1}{2\pi i}\int_{\gamma-i\infty}^{\gamma+i\infty} dp\,\tilde{g}(p)\int_0^\infty dr\, \mathrm{e}^{(p-s)r}\int_0^r dt\, Q(t)\,\mathrm{e}^{-pt} \\ &= \frac{1}{2\pi i}\int_{\gamma-i\infty}^{\gamma+i\infty} dp\,\tilde{g}(p)\left\{\left[\frac{\mathrm{e}^{(p-s)r}}{p-s}\int_0^r dt\, Q(t)\,\mathrm{e}^{-pt}\right]_{r=0}^{r=\infty}\right. \\ &\quad \left. -\int_0^\infty dr\,\frac{\mathrm{e}^{(p-s)r}}{p-s}Q(r)\,\mathrm{e}^{-pr}\right\}. \end{aligned} \tag{5.103}$$

Here, $\tilde{Q}(s)$ is defined by the following equation:

$$\tilde{Q}(s) = \int_0^\infty dt\, Q(t)\,\mathrm{e}^{ist}, \tag{5.104}$$

and we get

$$\left[\frac{\mathrm{e}^{(p-s)r}}{p-s}\int_0^r dt\, Q(t)\,\mathrm{e}^{-pt}\right]_{r=0}^{r=\infty} = \lim_{r\to\infty}\frac{\mathrm{e}^{(p-s)r}}{p-s}\tilde{Q}(ip) = 0. \tag{5.105}$$

Because, we can set $p = \gamma + i\alpha(\gamma < s,\ \alpha = \text{real})$, and we get $\lim_{r\to\infty}|\mathrm{e}^{(p-s)r}| = \lim_{r\to\infty}|\mathrm{e}^{-(s-\gamma)r}| = 0$. Substituting this expression and Eq. (5.104) into Eq. (5.103), and finally using the residue theorem to get

$$\begin{aligned} F_{\mathrm{dum}}^{(21)} &= \frac{1}{2\pi i}\int_{\gamma-i\infty}^{\gamma+i\infty} dp\,\tilde{g}(p)\left[-\frac{1}{p-s}\int_0^\infty dr\, Q(r)\,\mathrm{e}^{-sr}\right] \\ &= \frac{\tilde{Q}(is)}{2\pi i}\int_{\gamma+i\infty}^{\gamma-i\infty} dp\,\frac{\tilde{g}(p)}{p-s} \\ &= \frac{\tilde{Q}(is)}{2\pi i}\int_{\mathrm{C}} dp\,\frac{\tilde{g}(p)}{p-s} = \tilde{g}(s)\tilde{Q}(is). \end{aligned} \tag{5.106}$$

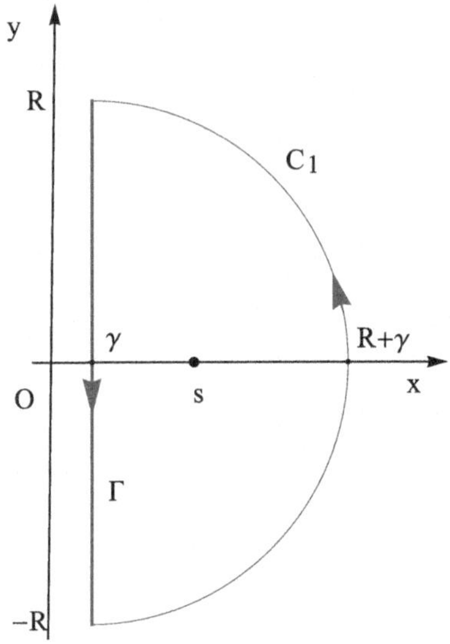

Figure 5.4 Single closed curve C. $\mathrm{C} = \Gamma + \mathrm{C}_1$ (C has a positive orientation). Straight line Γ: $x = \gamma, -R \leq y \leq R\ (R \to \infty)$, C_1: $x \geq \gamma$. A semicircle with radius R centered at point $(\gamma, 0)$ on the side of $x \geq \gamma$. The small black circle represents $(x, y) = (s, 0)$.

Here, the integration path C of Eq. (5.106) is a single closed curve on the complex plane $z = x + iy$ shown in the Fig. 5.4. Suppose that it consists of Γ and semicircle C_1. The vertical straight line Γ is $x = \gamma > 0, -R < y < R\ (R \to \infty)$, and the curve C_1 is a semicircle of radius R centered on $(\gamma, 0)$ and located on the $x \geq \gamma$ side. Also, $\widetilde{g}(p)$ is regular in the internal region including C, and is assumed to converge quickly to 0 on C_1 when $R \to \infty$.

Substituting the above result into Eq. (5.96), we get

$$\begin{aligned} 2\pi\widetilde{g}(s) &= \sum_{n=1}^{N} z_n D^{(n)} \frac{\mathrm{e}^{-(s+z_n)}}{s+z_n} + \frac{\mathrm{e}^{-s}[A + (A+B)s]}{s^2} \\ &\quad + 2\pi\rho \left[\widetilde{g}(s)\widetilde{Q}(is) - \sum_{n=1}^{N} D^{(n)} \widetilde{g}(z_n) \frac{\mathrm{e}^{-(z_n+s)}}{z_n+s} \right]. \\ \therefore\ 2\pi \left[1 - \rho\widetilde{Q}(is) \right] \widetilde{g}(s) & \\ = \mathrm{e}^{-s} & \left[\left(\frac{1}{s} + \frac{1}{s^2} \right) A + \frac{1}{s} B - \sum_{n=1}^{N} \frac{z_n \mathrm{e}^{-z_n}}{s+z_n} C^{(n)} \right]. \end{aligned} \tag{5.107}$$

Equation (5.107) is the Laplace transform of Baxter's second equation. From Eqs. (5.104), (5.87), and (5.94), we get

$$\begin{aligned}\widetilde{Q}(is) &= \int_0^1 dt\; Q^{(0)}(t)\,\mathrm{e}^{-st} + \int_0^\infty dt\; Q^{(1)}(t)\,\mathrm{e}^{-st}\\ &= \int_0^1 dt\left[\frac{1}{2}A(t^2-1) + B(t-1) + \sum_{n=1}^{N} C^{(n)}(\mathrm{e}^{-z_n t} - \mathrm{e}^{-z_n})\right]\mathrm{e}^{-st}\\ &\quad + \sum_{n=1}^{N} D^{(n)} \int_0^\infty dt\,\mathrm{e}^{-(z_n+s)t}\\ &= \psi_A(s)A + \psi_B(s)B + \sum_{n=1}^{N}\left[\psi_C^{(n)}(s)C^{(n)} + \psi_D^{(n)}(s)D^{(n)}\right],\end{aligned} \tag{5.108}$$

where,

$$\psi_A(s) = \frac{1}{s^3}\left[\left(1-\frac{s^2}{2}\right) - (1+s)\mathrm{e}^{-s}\right], \quad \psi_B(s) = \frac{1}{s^2}\left(1 - s - \mathrm{e}^{-s}\right),$$

$$\psi_C^{(n)}(s) = \frac{1}{s+z_n}\left[1 - \mathrm{e}^{-z_n} - \frac{z_n(1-\mathrm{e}^{-s})\mathrm{e}^{-z_n}}{s}\right], \quad \psi_D^{(n)}(s) = \frac{1}{s+z_n}.$$

5.6.4 Laplace Transform of Baxter's First Equation

Next, the Laplace transform of Eq. (5.10) is obtained. Under MSA, when the interparticle interaction $\phi(r)$ is given by multi-Yukawa terms, the direct correlation function $c(r)$ is given by

$$c(r) = \frac{1}{r}\sum_{n=1}^{N} K^{(n)}\mathrm{e}^{-z_n r}, \quad r > 1. \tag{5.109}$$

Substitute Eq. (5.109) for Eq. (5.10)

$$\begin{aligned}2\pi\sum_{n=1}^{N} K^{(n)}\mathrm{e}^{-z_n r} &= -\frac{dQ^{(1)}(r)}{dr} + \rho\int_0^\infty dt\; Q(t)\frac{dQ^{(1)}(r+t)}{dr}\\ &= \sum_{n=1}^{N} z_n D^{(n)}\mathrm{e}^{-z_n r} - \rho\sum_{n=1}^{N} z_n D^{(n)}\mathrm{e}^{-z_n r}\int_0^\infty dt\; Q(t)\mathrm{e}^{-z_n t}\\ &= \sum_{n=1}^{N} z_n D^{(n)}\mathrm{e}^{-z_n r} - \rho\sum_{n=1}^{N} z_n D^{(n)}\mathrm{e}^{-z_n r}\widetilde{Q}(iz_n)\\ &= \sum_{n=1}^{N} z_n D^{(n)}\mathrm{e}^{-z_n r}\left[1 - \rho\widetilde{Q}(iz_n)\right].\end{aligned} \tag{5.110}$$

From the above expression, the following expression is obtained from the independence of $\exp(-z_n r)$ $(n = 1, 2, \cdots, N)$:

$$2\pi K^{(n)} = z_n D^{(n)} \left[1 - \rho \widetilde{Q}(iz_n)\right]. \tag{5.111}$$

This is the Laplace transform of Baxter's first equation.

5.6.5 Analytical Solution of Baxter-Type OZ Equation

The Baxter function $Q(r)$ is given by Eqs. (5.87), (5.91), and (5.94). A and B included in Eq. (5.94) can be obtained from Eq. (5.82) (see Appendix D). Looking at the result, $Q(r)$ is found to be a function of $2N$ unknowns $\left(\vec{\widetilde{g}}, \vec{D}\right) = (\widetilde{g}(z_1), \widetilde{g}(z_2), \cdots, \widetilde{g}(z_N), D^{(1)}, D^{(2)}, \cdots, D^{(N)})$.

The Baxter-type OZ equations are obtained by placing $s = z_k$ $(1 \leq k \leq N)$ in Eqs. (5.107) and (5.111). The $2N$ equations include $2N$ unknown quantities $\left(\vec{\widetilde{g}}, \vec{D}\right)$. The unknowns are obtained by solving analytically the $2N$ equations under MSA for multi-Yukawa interaction. The analytical solutions can be easily applied to the SCOZA. This is shown below.

Comparing $c(r)$ given by Eq. (5.109) and $c(r)$ given by Eqs. (5.78), (5.79), and (3.2), we can get the following:

$$K^{(1)} = \mathcal{K}^{(1)} e^{z_1}, \quad K^{(n)} = \beta R(\rho, \beta) a_n e^{z_n} \ (2 \leq n \leq N),$$
$$\text{and } a_1 \equiv 2\pi \mathcal{K}^{(1)}. \tag{5.112}$$

Here, the unknown quantities $\left(\vec{\widetilde{g}}, \vec{D}\right)$ are normalized as follows:

$$\mathcal{G}_n = z_n e^{z_n} \widetilde{g}(z_n), \quad \mathcal{D}_n = \frac{z_n}{a_n e^{z_n}} D^{(n)} \quad (1 \leq n \leq N). \tag{5.113}$$

In the conventional SCOZA using $D^{(n)}$, the range of $D^{(n)}$ fluctuates exponentially, which cause various difficulties in performing numerical calculations and analysis. However, when using the newly introduced $\mathcal{D}_n$ in this book, the difficulty will be greatly improved because the domain will fit in a reasonable area. This is noteworthy.

If you use $\mathcal{G}_n$ of Eq. (5.113) for Eqs. (1.10), (3.2), and (5.89), the following formula is obtained:

$$\mathcal{F}_1 = u + 2\pi\rho^2 \sum_{n=2}^{N} \frac{a_n}{z_n} \mathcal{G}_n = 0. \tag{5.114}$$

When $n = 1$ is set in Eq. (5.111), the following formula is obtained:

$$\mathscr{F}_2 = 1 - \mathscr{D}_1 \left[1 - \rho\widetilde{Q}(iz_1)\right] = 0. \tag{5.115}$$

To erase the unknown $R(\rho, \beta)$ contained in $K^{(n)}$ $(2 \leq n \leq N)$ of Eq. (5.111), subtract each side of the nth expression from each side of the $(n+1)$th expression, the following expression is obtained:

$$\begin{aligned}\mathscr{F}_{n+1} = \mathscr{D}_n &\left[1 - \rho\widetilde{Q}(iz_n)\right] \\ &-\mathscr{D}_{n+1}\left[1 - \rho\widetilde{Q}(iz_{n+1})\right] = 0,\end{aligned} \tag{5.116}$$

where, $(2 \leq n \leq N-1)$. If we set $s = z_n$ in the Eq. (5.108), after a simple long calculation, $\widetilde{Q}(iz_n)$ is given by the following:

$$\widetilde{Q}(iz_n) = \hat{Q}_0(n) + \sum_{k=1}^{N} \left[\mathscr{Q}_1(n, k) + \mathscr{Q}_2(n, k)\mathscr{G}_k\right] \mathscr{P}_k, \tag{5.117}$$

where, $(1 \leq n \leq N)$. Furthermore, if $s = z_k$ is set in Eq. (5.107), the following equation is obtained after a long simple calculation:

$$\begin{aligned}\mathscr{F}_{N+k} &= F_0(k) + F_1(k)\mathscr{G}_k \\ &+\sum_{n=1}^{N}\left\{\mathscr{F}_2(k, n) + \mathscr{F}_3(k, n)\mathscr{G}_n + \mathscr{G}_k\left[\mathscr{F}_4(k, n) + \mathscr{F}_5(k, n)\mathscr{G}_n\right]\right\}\mathscr{D}_n \\ &= 0. \ (1 \leq k \leq N).\end{aligned} \tag{5.118}$$

The coefficients $F_0(k)$, $F_1(k)$, $\mathscr{F}_2(k, n)$, $\mathscr{F}_3(k, n)$, $\mathscr{F}_4(k, n)$, and $\mathscr{F}_5(k, n)$ are written in Appendix C. $2N$ unknowns $(\vec{\mathscr{G}}, \vec{\mathscr{D}}) = (\mathscr{G}_1, \mathscr{G}_2, \cdots, \mathscr{G}_N, \mathscr{D}_1, \mathscr{D}_2, \cdots, \mathscr{D}_N)$ can be obtained by solving $\mathscr{F}_n$ $(1 \leq n \leq 2N)$. Method to find u of Eq. (5.114) is explained in the next section.

5.7 MSA Analytic Display of Diffusion Equation for *u*

SCOZA's closure (5.78) includes a new unknown variable $R(\rho, \beta)$ that does not exist in MSA, so an equation is needed to determine it. It is a conditional expression (1.24) for the state equations derived from the energy equation and the compressibility equation to coincide. Since the expression contains $\chi_{\rm red}$ and u, we will display them using the MSA analytical solution obtained above.

From Eq. (5.15),

$$\widetilde{h}(0) = \lim_{q\to 0} 4\pi \int_0^\infty dr\, r^2 h(r) \frac{\sin(qr)}{qr} = 4\pi \int_0^\infty dr\, r^2 h(r). \quad (5.119)$$

Using Eqs. (5.119), (1.21), (1.7), (5.23), and (5.24),

$$\begin{aligned}\chi_{\rm red} &= \kappa\rho kT = 1 + 4\pi\rho \int_0^\infty h(r) r^2 dr = 1 + \rho\widetilde{h}(0) \\ &= \frac{1}{1-\rho\widetilde{c}(0)} = \frac{1}{Q^{\dagger 2}(0)}.\end{aligned} \quad (5.120)$$

By the way, from Eqs. (5.25) and (5.82),

$$Q^\dagger(0) = 1 - \rho \int_0^\infty dr\, Q(r) = \frac{A}{2\pi}. \quad (5.121)$$

From Eqs. (5.121) and (5.120),

$$\frac{1}{\chi_{\rm red}} = \left(\frac{A}{2\pi}\right)^2. \quad (5.122)$$

From Eq. (5.114),

$$u = -2\pi\rho^2 \sum_{n=2}^{N} \frac{a_n}{z_n} \mathscr{G}_n. \quad (5.123)$$

Therefore, the conditional expression (1.24) can be written as follows:

$$\rho \frac{\partial^2 u}{\partial \rho^2} = \mathcal{B}(\rho, u) \frac{\partial u}{\partial \beta}, \quad (5.124)$$

$$\begin{aligned}\mathcal{B}(\rho, u) &= \frac{\partial}{\partial u} \frac{1}{\chi_{\rm red}} = \frac{\partial}{\partial u} \left(\frac{A}{2\pi}\right)^2 = \frac{A}{2\pi^2} \frac{\partial A}{\partial u} \\ &= \frac{A}{2\pi^2} \sum_{n=1}^{2N} \frac{\partial A}{\partial x_n} \frac{\partial x_n}{\partial u},\end{aligned} \quad (5.125)$$

where, $x_n = \mathscr{G}_n$, $x_{N+n} = \mathscr{D}_n$ $(n = 1, 2, \cdots, N)$. The quantity A is given in Appendix D. How to find $\partial x_n / \partial u$ of Eq. (5.125) is shown in Appendix E.

5.7.1 Boundary Conditions

In the diffusion equation (or thermal conduction equation) for u in Eq. (5.124), the independent variable ρ corresponds to spatial coordinate, and β corresponds to time. Each variable of ρ and β is considered within the range of $0 < \rho < \rho_{\mathrm{J}}$ and $0 \leq \beta \leq \beta_f$. When $\rho = 0$, we can regard as $\phi(r) = 0$. Therefore, as a boundary condition, we get

$$u(\rho = 0, \beta) = 0, \qquad 0 \leq \beta \leq \beta_{\mathrm{f}}. \tag{5.126}$$

At high density $\rho = \rho_{\mathrm{J}}$, if we adopt the so-called high temperature approximation [Hansen and McDonald (1986)],

$$\frac{\partial^2 u}{\partial \rho^2}(\rho_{\mathrm{J}}, \beta) = \frac{\partial^2 u}{\partial \rho^2}(\rho_{\mathrm{J}}, \beta = 0), \qquad 0 \leq \beta \leq \beta_{\mathrm{f}}. \tag{5.127}$$

5.7.2 Initial Conditions

At $\beta = 0$ (or $T = \infty$), the interaction between particles is negligibly small compared to the thermal kinetic energy, so it can be set as $\phi(r) = 0$. In this case, we get $K^{(n)} = 0$ $(2 \leq n \leq N)$ from Eqs. (5.78), (5.79), and (5.109). Therefore, $D^{(n)} = \mathscr{D}_n = 0$ $(2 \leq n \leq N)$ are obtained from Eqs. (5.111) and (5.113). Therefore, the unknown quantities $(\mathscr{G}_1, \mathscr{D}_1)$ are obtained as solutions of the following equations obtained by placing $\mathscr{D}_n = 0$ $(2 \leq n \leq N)$ in Eqs. (5.115), (5.117), and (5.118).

$$\begin{aligned}\mathscr{F}_2 = 1 - \mathscr{D}_1\Big\langle 1 - \rho\Big[\hat{Q}_0(1) + \Big\{\mathscr{Q}_1(1, 1)\\ +\mathscr{Q}_2(1, 1)\mathscr{G}_1\Big\}\mathscr{D}_1\Big]\Big\rangle = 0.\end{aligned} \tag{5.128}$$

$$\begin{aligned}\mathscr{F}_{N+1} = F_0(1) + F_1(1)\mathscr{G}_1 + \Big\{\mathscr{F}_2(1, 1) + \mathscr{F}_3(1, 1)\mathscr{G}_1\\ +\mathscr{G}_1\left[\mathscr{F}_4(1, 1) + \mathscr{F}_5(1, 1)\mathscr{G}_1\right]\Big\}\mathscr{D}_1 = 0.\end{aligned} \tag{5.129}$$

The above two equations can be solved stably using the Newton–Raphson method as simultaneous equations of $\mathscr{G}_1$ and $\mathscr{D}_1$. Alternatively, solve Eq. (5.129) for $\mathscr{D}_1$, substitute it into Eq. (5.128), and rewrite it as an equation containing only $\mathscr{G}_1$. We can get $\mathscr{G}_1$ by solving this equation. However, this method is not recommended because it tends to cause problems in the stability of numerical calculations.

If $\mathscr{G}_1$ and $\mathscr{D}_1$ are obtained, $\mathscr{G}_k$ $(2 \leq k \leq N)$ are given by the following equations obtained from Eq. (5.118):

$$\mathscr{G}_k = -\frac{F_0(k) + [\mathscr{F}_2(k, 1) + \mathscr{F}_3(k, 1)\mathscr{G}_1]\,\mathscr{D}_1}{F_1(k) + [\mathscr{F}_4(k, 1) + \mathscr{F}_5(k, 1)\mathscr{G}_1]\,\mathscr{D}_1}, \qquad (5.130)$$

where, $(2 \leq k \leq N)$. The excess internal energy $u(\rho, \beta = 0)$ can be obtained using Eq. (5.123). Substituting this $u(\rho, 0)$ into Eq. (5.124) gives $u(\rho, \Delta\beta)$. Here, $\Delta\beta$ is a step size of β. For this new $u(\rho, \Delta\beta)$, if you solve $\mathscr{F}_n = 0$ $(1 \leq n \leq 2N)$, $(\vec{\mathscr{G}}, \vec{\mathscr{D}})$ are obtained for $\beta = \Delta\beta$ and ρ. By repeating the same, $u(\rho, \beta_j)$ for any j are obtained. Other physical quantities such as pressure and chemical potential can be derived from the thermodynamic potential $u(\rho, \beta_j)$.

5.7.3 The Spinodal Curve

As you can see from the diffusion equation (5.124) and Eq. (5.125) that gives the coefficient $\mathcal{B}$, when $\mathcal{B} > 0$ or $A > 0$, the solution of the diffusion equation is stable (see Appendix F). In the temperature range higher than the critical temperature kT_c $(= 1/\beta_c)$, the quantity A has positive value at any density ρ, so the fluid (gas) exists stably. However, if the temperature is lowered to the critical temperature T_c, the quantity A becomes 0 at the critical density ρ_c. When the temperature is further lowered from T_c, the quantity A becomes negative in the region $(\rho_{S1} \leq \rho \leq \rho_{S2})$ where $A(\rho_{S1}) = A(\rho_{S2}) = 0$, and the unstable region of $A < 0$ (gas–liquid coexistence area) expands. The curve that appears below the critical temperature and satisfies $A(\rho_{S1}) = A(\rho_{S2}) = 0$ is called a spinodal curve. Any physical quantities on this curve can be obtained from the solutions $\left(\vec{\mathscr{G}}, \vec{\mathscr{D}}\right)$ of the following equations:

$$A = \mathscr{F}_2 = \mathscr{F}_3 = \cdots = \mathscr{F}_{2N} = 0. \qquad (5.131)$$

Substitution of the solutions $\mathscr{G}_2, \mathscr{G}_3, \cdots, \mathscr{G}_N$ included in $\vec{\mathscr{G}}$ into Eq. (5.123) gives excess internal energy $u_S(\rho)$ on the spinodal curve. The boundary condition on the spinodal curve is expressed by the following equation according to [Pini (1998)]:

$$u(\rho_{S_i}, \beta) = u_S(\rho_{S_i}), \quad (i = 1, 2), \quad \rho_{S_1} < \rho_{S_2}. \qquad (5.132)$$

If $u(\rho, \beta)$ is obtained from Eq. (5.124), pressure p and chemical potential μ are derived, respectively, by integrating Eqs. (1.16) and

(1.20) with respect to β. Also, if Eqs. (1.21) and (1.23) obtained from the compressibility equation are integrated with respect to ρ, p and μ can be obtained. The quantities p and μ obtained from these two different routes match with high accuracy, respectively. Details are explained in Chapter 7.

5.8 Numerical Calculation Method of Diffusion (Heat Conduction) Type Equation

Equation (5.124), which u should satisfy, is solved numerically by learning from [Pini (1998)]. To do this, the step size of β at k step is set to $\Delta\beta_k$, and the value of β at n step is given by $\beta_n = \sum_{k=1}^{n} \Delta\beta_k$. In this section, the density increment $\Delta\rho$ is constant regardless of temperature and density. For $\Delta\beta$, because the temperature dependence of the diffusion coefficient $\mathcal{B}(\rho, u)$ is large in the vicinities of $\beta = 0$ and $\beta = \beta_c$, it is necessary to set a smaller value of $\Delta\beta$ in these regions than in the other temperature regions.

For any temperature β_n, if $u_j^n = u(\rho_j, \beta_n)$ are obtained for all j which satisfy $0 < \rho_j \leq \rho_J$, $u_j^{n+1/2}$ can be obtained as solutions of the following J simultaneous linear equations of predictor.

$$\rho_j \frac{u_{j+1}^{n+1/2} - 2u_j^{n+1/2} + u_{j-1}^{n+1/2}}{(\Delta\rho)^2} = 2\mathcal{B}(\rho_j, u_j^n)\frac{u_j^{n+1/2} - u_j^n}{\Delta\beta}. \quad (5.133)$$

If $u_j^{n+1/2}$ are obtained, u_j^{n+1} can be obtained from the next J correctors.

$$\frac{1}{2}\rho_j \left[\frac{u_{j+1}^{n+1} - 2u_j^{n+1} + u_{j-1}^{n+1}}{(\Delta\rho)^2} + \frac{u_{j+1}^{n} - 2u_j^{n} + u_{j-1}^{n}}{(\Delta\rho)^2}\right]$$
$$= \mathcal{B}(\rho_j, u_j^{n+1/2})\frac{u_j^{n+1} - u_j^n}{\Delta\beta}. \quad (5.134)$$

In this way, u_j^n $(0 < j \leq J)$ can be obtained for any n.

- Boundary conditions

The boundary conditions (5.126) and (5.127) can be written as follows:

At $\rho = 0$,

$$u_0^n = 0 \qquad \text{for any } \beta_n \geq 0. \quad (5.135)$$

At $\rho = \rho_J$,

$$u_J^n - 2u_{J-1}^n + u_{J-2}^n = u_J^0 - 2u_{J-1}^0 + u_{J-2}^0 = \Delta^2 u^0, \tag{5.136}$$

for any n.

5.8.1 In the Case of $T \geq T_c$

5.8.1.1 Predictor

Predictor is divided into the following two cases:

- for $1 \leq j \leq J-1$,

$$\frac{\rho_j}{(\Delta\rho)^2} u_{j-1}^{n+1/2} - 2\left[\frac{\rho_j}{(\Delta\rho)^2} + \frac{\mathcal{B}(\rho_j, u_j^n)}{\Delta\beta}\right] u_j^{n+1/2} + \frac{\rho_j}{(\Delta\rho)^2} u_{j+1}^{n+1/2}$$
$$= -2\frac{\mathcal{B}(\rho_j, u_j^n)}{\Delta\beta} u_j^n, \tag{5.137}$$

where, $u_0^{n+1/2} = 0$.

- for $j = J$,

$$u_{J-2}^{n+1/2} - 2u_{J-1}^{n+1/2} + u_J^{n+1/2} = \Delta^2 u^0. \tag{5.138}$$

5.8.1.2 Corrector

Corrector can be classified into the following two cases:

- for $1 \leq j \leq J-1$,

$$\frac{\rho_j}{(\Delta\rho)^2} u_{j-1}^{n+1} - 2\left[\frac{\rho_j}{(\Delta\rho)^2} + \frac{\mathcal{B}(\rho_j, u_j^{n+1/2})}{\Delta\beta}\right] u_j^{n+1} + \frac{\rho_j}{(\Delta\rho)^2} u_{j+1}^{n+1}$$
$$= -\frac{\rho_j}{(\Delta\rho)^2} u_{j+1}^n + 2\left[\frac{\rho_j}{(\Delta\rho)^2} - \frac{\mathcal{B}(\rho_j, u_j^{n+1/2})}{\Delta\beta}\right] u_j^n$$
$$-\frac{\rho_j}{(\Delta\rho)^2} u_{j-1}^n, \tag{5.139}$$

where, $u_0^n = u_0^{n+1} = 0$.

- for $j = J$,

$$u_{J-2}^{n+1} - 2u_{J-1}^{n+1} + u_J^{n+1} = \Delta^2 u^0. \tag{5.140}$$

5.8.1.3 Simultaneous linear equations for u_j $(1 \le j \le J)$

Both u_j^m $(1 \le j \le J)$ for the predictor and corrector described above are obtained as solutions of the following linear equations, where, $m = n + 1/2$ or $m = n + 1$.

$$\begin{bmatrix} b_1 & a_1 & 0 & 0 & 0 & 0 & 0 & 0 \\ a_2 & b_2 & a_2 & 0 & 0 & 0 & 0 & 0 \\ 0 & a_3 & b_3 & a_3 & 0 & 0 & 0 & 0 \\ \cdots \\ \cdots \\ 0 & 0 & 0 & 0 & 0 & a_{J-1} & b_{J-1} & a_{J-1} \\ 0 & 0 & 0 & 0 & 0 & 1 & -2 & 1 \end{bmatrix} \begin{bmatrix} u_1 \\ u_2 \\ u_3 \\ \cdots \\ \cdots \\ u_{J-1} \\ u_J \end{bmatrix} = \begin{bmatrix} r_1 \\ r_2 \\ r_3 \\ \cdots \\ \cdots \\ r_{J-1} \\ r_J \end{bmatrix}. \quad (5.141)$$

For the predictor, a_j, b_j, and r_j $(1 \le j \le J - 1)$ are given by the following equations:

$$a_j = \frac{\rho_j}{(\Delta\rho)^2} \quad (j = 1, 2, 3, \cdots, J - 1), \quad (5.142)$$

$$b_j = -2\left[\frac{\rho_j}{(\Delta\rho)^2} + \frac{\mathcal{B}(\rho_j, u_j^n)}{\Delta\beta}\right] \quad (j = 1, 2, \cdots, J - 1), \quad (5.143)$$

$$r_j = -2\frac{\mathcal{B}(\rho_j, u_j^n)}{\Delta\beta}u_j^n \quad (j = 1, 2, \cdots, J - 1), \quad r_J = \Delta^2 u^0. \quad (5.144)$$

For the corrector, those are given by

$$a_j = \frac{\rho_j}{(\Delta\rho)^2} \quad (j = 1, 2, 3, \cdots, J - 1), \quad (5.145)$$

$$b_j = -2\left[\frac{\rho_j}{(\Delta\rho)^2} + \frac{\mathcal{B}(\rho_j, u_j^{n+1/2})}{\Delta\beta}\right] \quad (j = 1, 2, \cdots, J - 1), \quad (5.146)$$

$$r_1 = -\frac{\rho_1}{(\Delta\rho)^2}u_2^n + 2\left[\frac{\rho_1}{(\Delta\rho)^2} - \frac{\mathcal{B}(\rho_1, u_1^{n+1/2})}{\Delta\beta}\right]u_1^n, \quad (5.147)$$

$$r_j = -\frac{\rho_j}{(\Delta\rho)^2}u_{j+1}^n + 2\left[\frac{\rho_j}{(\Delta\rho)^2} - \frac{\mathcal{B}(\rho_j, u_j^{n+1/2})}{\Delta\beta}\right]u_j^n - \frac{\rho_j}{(\Delta\rho)^2}u_{j-1}^n$$
$$(j = 2, 3, \cdots, J - 1), \quad (5.148)$$

$$r_J = \Delta^2 u^0. \quad (5.149)$$

5.8.1.4 Solution method of linear equations $A\mathbf{u} = \mathbf{r}$

Next, find the solution of simultaneous linear equations (5.141). However, for simplicity, the case of $J = 5$ will be described. It is easy to extend the obtained results for any J larger than 5.

5.8.1.5 LU decomposition of matrix A

Arbitrary square matrix A can be decomposed into $A = \hat{L}U$ using lower triangular matrix $\hat{L}$ and upper triangular matrix U. Here, the coefficient matrix A of Eq. (5.141) for $J = 5$ is decomposed into $A = \hat{L}U$. In this case, A, $\hat{L}$, and U can be written as follows:

$$A = \begin{bmatrix} b_1 & a_1 & 0 & 0 & 0 \\ a_2 & b_2 & a_2 & 0 & 0 \\ 0 & a_3 & b_3 & a_3 & 0 \\ 0 & 0 & a_4 & b_4 & a_4 \\ 0 & 0 & 1 & -2 & 1 \end{bmatrix}, \quad \hat{L} = \begin{bmatrix} L_{11} & 0 & 0 & 0 & 0 \\ L_{21} & L_{22} & 0 & 0 & 0 \\ L_{31} & L_{32} & L_{33} & 0 & 0 \\ L_{41} & L_{42} & L_{43} & L_{44} & 0 \\ L_{51} & L_{52} & L_{53} & L_{54} & L_{55} \end{bmatrix},$$

$$U = \begin{bmatrix} 1 & U_{12} & U_{13} & U_{14} & U_{15} \\ 0 & 1 & U_{23} & U_{24} & U_{25} \\ 0 & 0 & 1 & U_{34} & U_{35} \\ 0 & 0 & 0 & 1 & U_{45} \\ 0 & 0 & 0 & 0 & 1 \end{bmatrix}.$$

The following results are obtained from the relational expression $A_{ij} = \sum_{k=1}^{J} L_{ik}U_{kj}$ between the components of A, $\hat{L}$, and U:

$$A_{11} = b_1 = \sum_{k=1}^{J} L_{1k}U_{k1} = L_{11},$$

$$A_{12} = a_1 = \sum_{k=1}^{J} L_{1k}U_{k2} = L_{11}U_{12} = b_1U_{12}, \quad \therefore\ U_{12} = \frac{a_1}{b_1},$$

$$A_{1j} = 0 = \sum_{k=1}^{J} L_{1k}U_{kj} = b_1U_{1j}, \quad \therefore\ U_{1j} = 0\ (3 \le j \le 5).$$

If the same calculations are continued for other A_{ij}, the following results are obtained:

$$\hat{L} = \begin{bmatrix} L_{11} & 0 & 0 & 0 & 0 \\ L_{21} & L_{22} & 0 & 0 & 0 \\ 0 & L_{32} & L_{33} & 0 & 0 \\ 0 & 0 & L_{43} & L_{44} & 0 \\ 0 & 0 & 1 & L_{54} & L_{55} \end{bmatrix}, \quad U = \begin{bmatrix} 1 & U_{12} & 0 & 0 & 0 \\ 0 & 1 & U_{23} & 0 & 0 \\ 0 & 0 & 1 & U_{34} & 0 \\ 0 & 0 & 0 & 1 & U_{45} \\ 0 & 0 & 0 & 0 & 1 \end{bmatrix}, \quad (5.150)$$

where,

$$\begin{cases} U_{12} = \dfrac{a_1}{b_1}, \quad U_{i,i+1} = \dfrac{a_i}{b_i - U_{i-1,i}a_i} \quad (2 \leq i \leq J-1), \\ L_{11} = b_1, \quad L_{i,i-1} = a_i \quad (2 \leq i \leq J-1), \\ L_{ii} = b_i - U_{i-1,i}a_i \quad (2 \leq i \leq J-1), \\ L_{J,J-2} = 1, \quad L_{J,J-1} = -2 - U_{J-2,J-1}, \\ L_{JJ} = 1 + (2 + U_{J-2,J-1})U_{J-1,J}. \end{cases} \tag{5.151}$$

If transpose matrices of $\mathbf{u}$ and $\mathbf{r}$ are represented, respectively, by $\mathbf{u}^{\mathrm{t}} = (u_1, u_2, \cdots, u_5)$ and $\mathbf{r}^{\mathrm{t}} = (r_1, r_2, \cdots, r_5)$, the equations to be solved can be written as $A\mathbf{u} = \mathbf{r}$.

5.8.1.6 Solution of $A\mathbf{u} = \mathbf{r}$

$U\mathbf{u} = \mathbf{h}$ means $A\mathbf{u} = \hat{L}U\mathbf{u} = \hat{L}\mathbf{h} - \mathbf{r}$. So, we get

$$\hat{L}\mathbf{h} = \begin{bmatrix} L_{11}h_1 \\ L_{21}h_1 + L_{22}h_2 \\ L_{32}h_2 + L_{33}h_3 \\ L_{43}h_3 + L_{44}h_4 \\ h_3 + L_{54}h_4 + L_{55}h_5 \end{bmatrix} = \begin{bmatrix} r_1 \\ r_2 \\ r_3 \\ r_4 \\ r_5 \end{bmatrix}, \quad \text{where } \mathbf{h} = \begin{bmatrix} h_1 \\ h_2 \\ h_3 \\ h_4 \\ h_5 \end{bmatrix}.$$

Therefore,

$$\begin{aligned} h_1 &= \frac{r_1}{L_{11}}, \quad h_j = \frac{r_j - L_{j,j-1}h_{j-1}}{L_{jj}} \quad (2 \leq j \leq J-1), \\ h_J &= \frac{r_J - h_{J-2} - L_{J,J-1}h_{J-1}}{L_{JJ}}. \end{aligned} \tag{5.152}$$

And, from $U\mathbf{u} = \mathbf{h}$,

$$\begin{bmatrix} u_1 + U_{12}u_2 \\ u_2 + U_{23}u_3 \\ u_3 + U_{34}u_4 \\ u_4 + U_{45}u_5 \\ u_5 \end{bmatrix} = \begin{bmatrix} h_1 \\ h_2 \\ h_3 \\ h_4 \\ h_5 \end{bmatrix}.$$

We can get

$$u_J = h_J, \quad u_j = h_j - U_{j,j+1}u_{j+1} \quad (j = J-1, \cdots, 2, 1). \tag{5.153}$$

Here, if $p_j = U_{j,j+1}$ is set, the following equations are finally obtained:

$$u_J = h_J, \quad u_j = h_j - p_j u_{j+1} \quad (j = J-1, \cdots, 2, 1). \tag{5.154}$$

where,

$$\begin{cases} p_1 = \dfrac{a_1}{b_1}, \quad p_j = \dfrac{a_j}{b_j - a_j p_{j-1}} & (j = 2, 3, \cdots, J-1). \\ h_1 = \dfrac{r_1}{b_1}, \quad h_j = \dfrac{r_j - a_j h_{j-1}}{b_j - a_j p_{j-1}} & (j = 2, 3, \cdots, J-1). \\ h_J = \dfrac{r_J - h_{J-2} + (2 + p_{J-2}) h_{J-1}}{1 + (2 + p_{J-2}) p_{J-1}}. \end{cases}$$

As shown in Eq. (5.154), u_j is initially determined for $j = J$, and the value is determined in order of decreasing j.

Here, for the sake of simplicity, the case of $J = 5$ has been explained, but Eqs. (5.154) are expressions that hold for any J with $J > 5$.

5.8.2 In the Case of $T < T_c$ and $0 < \rho \leq \rho_L$

When the temperature reaches the critical temperature T_c with cooling, $A(\rho_c, \beta_c) = 0$ at the critical density ρ_c, and at other densities $\rho \neq \rho_c$, $A(\rho, \beta_c) > 0$. When the temperature drops further and $T < T_c$, there are two integers L and R, and the following equation holds:

$$\begin{cases} A(\rho_j, \beta) \leq 0 & \text{for} \quad L < j < R, \\ A(\rho_j, \beta) > 0 & \text{otherwise.} \end{cases} \tag{5.155}$$

In the region of $0 < \rho \leq \rho_L$, the fluid exists as a gas. In this case, $u(\rho_{L+1}, \beta) = u_S(\rho_{L+1})$ is obtained by substituting $\mathcal{G}_n$ $(2 \leq n \leq N)$ of the solutions of the boundary conditions (5.131) into Eq. (5.123).

5.8.2.1 Predictor

Predictor is classified into the following two types:

- $1 \leq j \leq L - 1$

$$\frac{\rho_j}{(\Delta\rho)^2} u_{j-1}^{n+1/2} - 2\left[\frac{\rho_j}{(\Delta\rho)^2} + \frac{\mathcal{B}(\rho_j, u_j^n)}{\Delta\beta}\right] u_j^{n+1/2} + \frac{\rho_j}{(\Delta\rho)^2} u_{j+1}^{n+1/2}$$
$$= -2\frac{\mathcal{B}(\rho_j, u_j^n)}{\Delta\beta} u_j^n. \tag{5.156}$$

- $j = L$

$$\frac{\rho_L}{(\Delta\rho)^2} u_{L-1}^{n+1/2} - 2\left[\frac{\rho_L}{(\Delta\rho)^2} + \frac{\mathcal{B}(\rho_L, u_L^n)}{\Delta\beta}\right] u_L^{n+1/2}$$
$$= -2\frac{\mathcal{B}(\rho_L, u_L^n)}{\Delta\beta} u_L^n - \frac{\rho_L}{(\Delta\rho)^2} u_S(\rho_{L+1}). \quad (5.157)$$

Here, $u_{L+1}^{n+1/2} = u_S(\rho_{L+1})$ from the boundary condition (5.132).

5.8.2.2 Corrector

Corrector is classified into the following two types:

- $1 \le j \le L-1$

$$\frac{\rho_j}{(\Delta\rho)^2} u_{j-1}^{n+1} - 2\left[\frac{\rho_j}{(\Delta\rho)^2} + \frac{\mathcal{B}(\rho_j, u_j^{n+1/2})}{\Delta\beta}\right] u_j^{n+1} + \frac{\rho_j}{(\Delta\rho)^2} u_{j+1}^{n+1}$$
$$= -\frac{\rho_j}{(\Delta\rho)^2} u_{j+1}^n + 2\left[\frac{\rho_j}{(\Delta\rho)^2} - \frac{\mathcal{B}(\rho_j, u_j^{n+1/2})}{\Delta\beta}\right] u_j^n$$
$$-\frac{\rho_j}{(\Delta\rho)^2} u_{j-1}^n. \quad (5.158)$$

- $j = L$

$$\frac{\rho_G}{(\Delta\rho)^2} u_{L-1}^{n+1} - 2\left[\frac{\rho_G}{(\Delta\rho)^2} + \frac{\mathcal{B}(\rho_G, u_L^{n+1/2})}{\Delta\beta}\right] u_L^{n+1}$$
$$= -2\frac{\rho_G}{(\Delta\rho)^2} u_S(\rho_{L+1}) + 2\left[\frac{\rho_G}{(\Delta\rho)^2} - \frac{\mathcal{B}(\rho_G, u_L^{n+1/2})}{\Delta\beta}\right] u_L^n$$
$$-\frac{\rho_G}{(\Delta\rho)^2} u_{L-1}^n. \quad (5.159)$$

Here, from the boundary condition (5.132), $u_{L+1}^n = u_{L+1}^{n+1} = u_S(\rho_{L+1})$.

5.8.2.3 A system of linear equations for u_j $(1 \le j \le L)$

The simultaneous linear equations of u_j $(1 \le j \le L)$ included in the predictor and the collector are given by the following equations:

$$\begin{bmatrix} b_1 & a_1 & 0 & 0 & 0 \\ a_2 & b_2 & a_2 & 0 & 0 \\ 0 & a_3 & b_3 & a_3 & 0 \\ 0 & 0 & a_4 & b_4 & a_4 \\ 0 & 0 & 0 & a_5 & b_5 \end{bmatrix} \begin{bmatrix} u_1 \\ u_2 \\ u_3 \\ u_4 \\ u_5 \end{bmatrix} = \begin{bmatrix} r_1 \\ r_2 \\ r_3 \\ r_4 \\ r_5 \end{bmatrix}. \quad (5.160)$$

Here, for simplicity, $L = 5$ is set. It is easy to extend to any L that makes $L > 5$.

Any coefficient matrix A can be decomposed into $A = \hat{L}U$ using the lower triangular matrix $\hat{L}$ and the upper triangular matrix U. Because $A_{ij} = \sum_{k=1}^{L} L_{ik}U_{kj}$, $\hat{L}$ and U can be written as follows:

$$\hat{L} = \begin{bmatrix} L_{11} & 0 & 0 & 0 & 0 \\ L_{21} & L_{22} & 0 & 0 & 0 \\ 0 & L_{32} & L_{33} & 0 & 0 \\ 0 & 0 & L_{43} & L_{44} & 0 \\ 0 & 0 & 0 & L_{54} & L_{55} \end{bmatrix}, \quad U = \begin{bmatrix} 1 & U_{12} & 0 & 0 & 0 \\ 0 & 1 & U_{23} & 0 & 0 \\ 0 & 0 & 1 & U_{34} & 0 \\ 0 & 0 & 0 & 1 & U_{45} \\ 0 & 0 & 0 & 0 & 1 \end{bmatrix}, \tag{5.161}$$

where,

$$L_{11} = b_1, \quad L_{ii} = b_i - a_i \frac{a_{i-1}}{L_{i-1,i-1}}, \quad L_{i,i-1} = a_i \quad (2 \le i \le L),$$
$$U_{i,i+1} = \frac{a_i}{L_{ii}} \quad (1 \le i \le L-1). \tag{5.162}$$

5.8.2.4 Solutions of $A\mathbf{u} = \mathbf{r}$

$U\mathbf{u} = \mathbf{h}$ means $A\mathbf{u} = \hat{L}U\mathbf{u} = \hat{L}\mathbf{h} = \mathbf{r}$, and we get

$$\hat{L}\mathbf{h} = \begin{bmatrix} L_{11}h_1 \\ L_{21}h_1 + L_{22}h_2 \\ L_{32}h_2 + L_{33}h_3 \\ L_{43}h_3 + L_{44}h_4 \\ L_{54}h_4 + L_{55}h_5 \end{bmatrix} = \begin{bmatrix} r_1 \\ r_2 \\ r_3 \\ r_4 \\ r_5 \end{bmatrix}, \quad \text{where } \mathbf{h} = \begin{bmatrix} h_1 \\ h_2 \\ h_3 \\ h_4 \\ h_5 \end{bmatrix}.$$

Therefore,

$$h_1 = \frac{r_1}{L_{11}}, \quad h_j = \frac{r_j - L_{j,j-1}h_{j-1}}{L_{jj}} \quad (2 \le j \le L), \tag{5.163}$$

and, from $U\mathbf{u} = \mathbf{h}$,

$$\begin{bmatrix} u_1 + U_{12}u_2 \\ u_2 + U_{23}u_3 \\ u_3 + U_{34}u_4 \\ u_4 + U_{45}u_5 \\ u_5 \end{bmatrix} = \begin{bmatrix} h_1 \\ h_2 \\ h_3 \\ h_4 \\ h_5 \end{bmatrix}.$$

Therefore,

$$u_L = h_L, \quad u_j = h_j - U_{j,j+1}u_{j+1} \quad (j = L-1, \cdots, 2, 1). \tag{5.164}$$

Here, when $p_j = U_{j,j+1}$ is set, the following equations are finally obtained:

$$u_L = h_L, \quad u_j = h_j - p_j u_{j+1} \quad (j = L-1, L-2, \cdots, 1). \quad (5.165)$$

where,

$$\begin{cases} p_1 = \dfrac{a_1}{b_1}, & p_j = \dfrac{a_j}{b_j - a_j p_{j-1}} \quad (j = 2, 3, \cdots, L-1). \\ h_1 = \dfrac{r_1}{b_1}, & h_j = \dfrac{r_j - a_j h_{j-1}}{b_j - a_j p_{j-1}} \quad (j = 2, 3, \cdots, L). \end{cases}$$

Here, $L = 5$, but Eq. (5.165) are expressions that hold for any L with $L > 5$. Here, coefficients a_j, b_j, and r_j $(1 \leq j \leq L)$ for Predictor are given by the following equations:

$$a_j = \frac{\rho_j}{(\Delta\rho)^2} \quad (j = 1, 2, 3, \cdots, L),$$

$$b_j = -2\left[\frac{\rho_j}{(\Delta\rho)^2} + \frac{\mathcal{B}(\rho_j, u_j^n)}{\Delta\beta}\right] \quad (j = 1, 2, \cdots, L),$$

$$r_j = -2\frac{\mathcal{B}(\rho_j, u_j^n)}{\Delta\beta}u_j^n \quad (j = 1, 2, \cdots, L-1),$$

$$r_L = -2\frac{\mathcal{B}(\rho_G, u_L^n)}{\Delta\beta}u_L^n - \frac{\rho_L}{(\Delta\rho)^2}u_S(\rho_{L+1}).$$

Also, for Collector, those are given by the following equations:

$$a_j = \frac{\rho_j}{(\Delta\rho)^2} \quad (j = 1, 2, 3, \cdots, L),$$

$$b_j = -2\left[\frac{\rho_j}{(\Delta\rho)^2} + \frac{\mathcal{B}(\rho_j, u_j^{n+1/2})}{\Delta\beta}\right] \quad (j = 1, 2, \cdots, L),$$

$$r_1 = -\frac{\rho_1}{(\Delta\rho)^2}u_2^n + 2\left[\frac{\rho_1}{(\Delta\rho)^2} - \frac{\mathcal{B}(\rho_1, u_1^{n+1/2})}{\Delta\beta}\right]u_1^n,$$

$$r_j = -\frac{\rho_j}{(\Delta\rho)^2}u_{j+1}^n + 2\left[\frac{\rho_j}{(\Delta\rho)^2} - \frac{\mathcal{B}(\rho_j, u_j^{n+1/2})}{\Delta\beta}\right]u_j^n - \frac{\rho_j}{(\Delta\rho)^2}u_{j-1}^n$$
$$(j = 2, 3, \cdots, L-1),$$

$$r_L = -2\frac{\rho_G}{(\Delta\rho)^2}u_S(\rho_{L+1}) + 2\left[\frac{\rho_G}{(\Delta\rho)^2} - \frac{\mathcal{B}(\rho_G, u_L^{n+1/2})}{\Delta\beta}\right]u_L^n$$
$$-\frac{\rho_G}{(\Delta\rho)^2}u_{L-1}^n.$$

5.8.3 In the Case of $T < T_c$ and $\rho_R \leq \rho \leq \rho_J$

When $T < T_c$, the fluid exists as a liquid in the region of $\rho_R \leq \rho \leq \rho_J$. In this case, $u(\rho_{R-1}, \beta) = u_S(\rho_{R-1})$ is obtained from the boundary condition.

5.8.3.1 Predictor

Predictor is classified into the following three types:

- In the case of $j = R$,

$$-2\left[\frac{\rho_R}{(\Delta\rho)^2} + \frac{\mathcal{B}(\rho_R, u_R^n)}{\Delta\beta}\right] u_R^{n+1/2} + \frac{\rho_R}{(\Delta\rho)^2} u_{R+1}^{n+1/2}$$
$$= -2\frac{\mathcal{B}(\rho_R, u_R^n)}{\Delta\beta} u_R^n - \frac{\rho_R}{(\Delta\rho)^2} u_S(\rho_{R-1}). \tag{5.166}$$

- In the case of $R + 1 \leq j \leq J - 1$,

$$\frac{\rho_j}{(\Delta\rho)^2} u_{j-1}^{n+1/2} - 2\left[\frac{\rho_j}{(\Delta\rho)^2} + \frac{\mathcal{B}(\rho_j, u_j^n)}{\Delta\beta}\right] u_j^{n+1/2} + \frac{\rho_j}{(\Delta\rho)^2} u_{j+1}^{n+1/2}$$
$$= -2\frac{\mathcal{B}(\rho_j, u_j^n)}{\Delta\beta} u_j^n. \tag{5.167}$$

- In the case of $j = J$,

$$u_{J-2}^{n+1/2} - 2u_{J-1}^{n+1/2} + u_J^{n+1/2} = \Delta^2 u^0. \tag{5.168}$$

5.8.3.2 Corrector

Corrector is classified into the following three categories:

- When $j = R$,

$$-2\left[\frac{\rho_R}{(\Delta\rho)^2} + \frac{\mathcal{B}(\rho_R, u_R^{n+1/2})}{\Delta\beta}\right] u_R^{n+1} + \frac{\rho_R}{(\Delta\rho)^2} u_{R+1}^{n+1}$$
$$= -\frac{\rho_R}{(\Delta\rho)^2} u_{R+1}^n + 2\left[\frac{\rho_R}{(\Delta\rho)^2} - \frac{\mathcal{B}(\rho_R, u_R^{n+1/2})}{\Delta\beta}\right] u_R^n$$
$$-2\frac{\rho_R}{(\Delta\rho)^2} u_S(\rho_{R-1}). \tag{5.169}$$

- In the case of $R+1 \leq j \leq J-1$,

$$\frac{\rho_j}{(\Delta\rho)^2}u_{j-1}^{n+1} - 2\left[\frac{\rho_j}{(\Delta\rho)^2} + \frac{\mathcal{B}(\rho_j, u_j^{n+1/2})}{\Delta\beta}\right]u_j^{n+1} + \frac{\rho_j}{(\Delta\rho)^2}u_{j+1}^{n+1}$$
$$= -\frac{\rho_j}{(\Delta\rho)^2}u_{j+1}^{n} + 2\left[\frac{\rho_j}{(\Delta\rho)^2} - \frac{\mathcal{B}(\rho_j, u_j^{n+1/2})}{\Delta\beta}\right]u_j^{n}$$
$$-\frac{\rho_j}{(\Delta\rho)^2}u_{j-1}^{n} \qquad (5.170)$$

- When $j = J$,

$$u_{J-2}^{n+1} - 2u_{J-1}^{n+1} + u_J^{n+1} = \Delta^2 u^0. \qquad (5.171)$$

5.8.3.3 A system of linear equations for $\rho_R \leq \rho \leq \rho_J$

The simultaneous linear equations to be solved are given by the following equations:

$$\begin{bmatrix} b_R & a_R & 0 & \cdots & 0 & 0 & 0 \\ a_{R+1} & b_{R+1} & a_{R+1} & \cdots & 0 & 0 & 0 \\ \cdots & & & & & & \\ \cdots & & & & & & \\ 0 & 0 & 0 & \cdots & a_{J-1} & b_{J-1} & a_{J-1} \\ 0 & 0 & 0 & \cdots & 1 & -2 & 1 \end{bmatrix} \begin{bmatrix} u_R \\ u_{R+1} \\ \cdots \\ \cdots \\ u_{J-1} \\ u_J \end{bmatrix} = \begin{bmatrix} r_R \\ r_{R+1} \\ \cdots \\ \cdots \\ r_{J-1} \\ r_J \end{bmatrix}. \qquad (5.172)$$

This has the same form as Eq. (5.141) where $j = 1$ is moved to $j = R$. Therefore, the solution u_j $(R \leq j \leq J)$ is given by the following equation:

$$u_J = h_J, \quad u_j = h_j - p_j u_{j+1} \quad (j = J-1, J-2, \cdots, R), \qquad (5.173)$$

where,

$$\begin{cases} p_R = \dfrac{a_R}{b_R}, \quad p_j = \dfrac{a_j}{b_j - a_j p_{j-1}} \quad (j = R+1, \cdots, J-1). \\ h_R = \dfrac{r_R}{b_R}, \quad h_j = \dfrac{r_j - a_j h_{j-1}}{b_j - a_j p_{j-1}} \quad (j = R+1, \cdots, J-1). \\ h_J = \dfrac{r_J - h_{J-2} + (2 + p_{J-2})h_{J-1}}{1 + (2 + p_{J-2})p_{J-1}}. \end{cases} \qquad (5.174)$$

Here, the coefficients for Predictor are given by the following equations:

$$a_j = \frac{\rho_j}{(\Delta\rho)^2} \quad (j = R, \cdots, J-1).$$

$$b_j = -2\left[\frac{\rho_j}{(\Delta\rho)^2} + \frac{\mathcal{B}(\rho_j, u_j^n)}{\Delta\beta}\right] \quad (j = R, \cdots, J-1).$$

$$r_R = -2\frac{\mathcal{B}(\rho_R, u_R^n)}{\Delta\beta}u_R^n - \frac{\rho_R}{(\Delta\rho)^2}u_S(\rho_{R-1}).$$

$$r_j = -2\frac{\mathcal{B}(\rho_j, u_j^n)}{\Delta\beta}u_j^n \quad (j = R+1, \cdots, J-1).$$

$$r_J = \Delta^2 u^0.$$

Also, the coefficients for Collector are given by the following equations:

$$a_j = \frac{\rho_j}{(\Delta\rho)^2} \quad (j = R, \cdots, J-1).$$

$$b_j = -2\left[\frac{\rho_j}{(\Delta\rho)^2} + \frac{\mathcal{B}(\rho_j, u_j^{n+1/2})}{\Delta\beta}\right] \quad (j = R, \cdots, J-1).$$

$$r_R = -\frac{\rho_R}{(\Delta\rho)^2}u_{R+1}^n + 2\left[\frac{\rho_R}{(\Delta\rho)^2} - \frac{\mathcal{B}(\rho_R, u_R^{n+1/2})}{\Delta\beta}\right]u_R^n$$
$$-2\frac{\rho_R}{(\Delta\rho)^2}u_S(\rho_{R-1}).$$

$$r_j = -\frac{\rho_j}{(\Delta\rho)^2}u_{j+1}^n + 2\left[\frac{\rho_j}{(\Delta\rho)^2} - \frac{\mathcal{B}(\rho_j, u_j^{n+1/2})}{\Delta\beta}\right]u_j^n - \frac{\rho_j}{(\Delta\rho)^2}u_{j-1}^n$$
$$(j = R+1, \cdots, J-1).$$

$$r_J = \Delta^2 u^0.$$

5.9 Numerical Calculation Method with Two Different Step Sizes of $\Delta\rho_a$ and $\Delta\rho_b$

Section 5.8 explains the numerical calculation method when the step size of density $\Delta\rho$ is constant. In order to increase the calculation accuracy efficiently, it may be better to change $\Delta\rho$. Here, such a case is described.

5.10 Method to Mark Density

The solution of the diffusion equation (5.124) is stable when the diffusion coefficient $\mathcal{B}(\rho, u)$ is positive. Therefore, calculation can be executed only in the region of $\mathcal{B}(\rho, u) > 0$. However, when $\beta = 0$, $\mathcal{B}(\rho, u)$ becomes negative when ρ exceeds a certain ρ_{J}. Since such J cannot be known in advance, J is set to a large value such that $\mathcal{B}(\rho_{\mathrm{J}}, u) < 0$ in the first trial calculation. It is necessary to finally determine the value of J such that $\mathcal{B}(\rho_{\mathrm{J}}, u) > 0$ and $\mathcal{B}(\rho_{\mathrm{J}+1}, u) \leq 0$ after a few trial estimations. The value of J determined after such trial calculations is shown in Table 5.1 for the intermolecular interaction $\phi_{\mathrm{c1}}(r)$.

Table 5.1 Method to cut density

J	j_{a}	ρ_{a}	$\Delta\rho_{\mathrm{a}}$	$\Delta\rho_{\mathrm{b}}$
7586	100	10^{-4}	10^{-6}	10^{-4}

Since the number of steps of density is limited due to computer performance, it is necessary to cut the density efficiently. In order to accurately determine the vaporization and liquefaction points in the low-temperature region, the density step size in the low-density region must be much smaller than the step size in the high-density region (see Chapter 7). Therefore, the density ρ is divided into two regions and engraved as follows (see Table 5.1):

(a) Region I $(0 < \rho \leq \rho_a)$

$$\Delta\rho_a = \frac{\rho_a}{j_a}, \quad \rho_j = j\Delta\rho_a \quad (j = 1, 2, 3, \cdots, j_a). \tag{5.175}$$

(b) Region II $(\rho_a < \rho \leq \rho_{\mathrm{J}})$

$$\Delta\rho_b = \frac{\rho_{\mathrm{J}} - \rho_a}{J - j_a}, \quad \rho_{j_a+j} = \rho_a + j\Delta\rho_b \quad (j = 1, 2, 3, \cdots, J - j_a). \tag{5.176}$$

5.10.1 In the Case of $T \geq T_{\mathrm{c}}$

5.10.1.1 Predictor

Predictor Eq. (5.133) can be classified as follows:

- In the case of $1 \leq j \leq j_a - 1$,

$$\frac{\rho_j}{(\Delta\rho_a)^2}u_{j-1}^{n+1/2} - 2\left[\frac{\rho_j}{(\Delta\rho_a)^2} + \frac{\mathcal{B}(\rho_j, u_j^n)}{\Delta\beta}\right]u_j^{n+1/2} + \frac{\rho_j}{(\Delta\rho_a)^2}u_{j+1}^{n+1/2} = -2\frac{\mathcal{B}(\rho_j, u_j^n)}{\Delta\beta}u_j^n. \quad (5.177)$$

- In the case of $j_a + 1 \leq j \leq J - 1$,

$$\frac{\rho_j}{(\Delta\rho_b)^2}u_{j-1}^{n+1/2} - 2\left[\frac{\rho_j}{(\Delta\rho_b)^2} + \frac{\mathcal{B}(\rho_j, u_j^n)}{\Delta\beta}\right]u_j^{n+1/2} + \frac{\rho_j}{(\Delta\rho_b)^2}u_{j+1}^{n+1/2} = -2\frac{\mathcal{B}(\rho_j, u_j^n)}{\Delta\beta}u_j^n. \quad (5.178)$$

- When $j = J$, from the boundary condition Eq. (5.127)

$$u_J^{n+1/2} - 2u_{J-1}^{n+1/2} + u_{J-2}^{n+1/2} = \Delta^2 u^0. \quad (5.179)$$

- In the case of $j = j_a$,

$$\rho_j\frac{\widetilde{u}_{j+1}^{n+1/2} - 2u_j^{n+1/2} + u_{j-1}^{n+1/2}}{(\Delta\rho_a)^2} = 2\mathcal{B}(\rho_j, u_j^n)\frac{u_j^{n+1/2} - u_j^n}{\Delta\beta}. \quad (5.180)$$

The density step changes from $\Delta\rho_a$ to $\Delta\rho_b$ at the point ρ_a, and this should be reflected in the difference equation of Eq. (5.180). For this purpose, a point $\hat{\rho}_x = \rho_a + \Delta\rho_a$ is virtually set, and the excess internal energy at that point is set to $\widetilde{u}_{j_a+1}^{n+1/2}$ (see Fig. 5.5). Method to find the value is explained in Section 5.10.1.2.

5.10.1.2 Interpolation of $\widetilde{u}_{j_a+1}^m$

For any temperature β_m, using excess internal energies $\hat{u}_1 = u(\hat{\rho}_1, \beta_m)$, $\hat{u}_2 = u(\hat{\rho}_2, \beta_m)$, and $\hat{u}_3 = u(\hat{\rho}_3, \beta_m)$, respectively, at 3 points of $\hat{\rho}_1 = \rho_a$, $\hat{\rho}_2 = \rho_a + \Delta\rho_b$, and $\hat{\rho}_3 = \rho_a + 2\Delta\rho_b$, the value of $\widetilde{u}_{j_a+1}^m$ is interpolated. In that case, the second order polynomial about ρ given by the following equation is used:

$$u(\rho, \beta_m) = c_2\rho^2 + c_1\rho + c_0. \quad (5.181)$$

The coefficients c_j $(j = 0, 1, 2)$ are given as solutions of the three simultaneous equations of $u(\hat{\rho}_j, \beta_m) = \hat{u}_j$ $(j = 1, 2, 3)$. When the obtained results are arranged focusing on $\hat{u}_j$ $(j = 1, 2, 3)$, $\widetilde{u}^m_{j_a+1}$ is given by the following equations:

$$\widetilde{u}^m_{j_a+1} = u(\hat{\rho}_x, \beta_m) = \hat{c}_{u1}\hat{u}_1 + \hat{c}_{u2}\hat{u}_2 + \hat{c}_{u3}\hat{u}_3, \tag{5.182}$$

where,

$$\hat{c}_{u1} = \frac{(\hat{\rho}_x - \hat{\rho}_2)(\hat{\rho}_x - \hat{\rho}_3)}{(\hat{\rho}_1 - \hat{\rho}_2)(\hat{\rho}_1 - \hat{\rho}_3)},$$

$$\hat{c}_{u2} = \frac{(\hat{\rho}_x - \hat{\rho}_1)(\hat{\rho}_x - \hat{\rho}_3)}{(\hat{\rho}_2 - \hat{\rho}_1)(\hat{\rho}_2 - \hat{\rho}_3)},$$

$$\hat{c}_{u3} = \frac{(\hat{\rho}_x - \hat{\rho}_1)(\hat{\rho}_x - \hat{\rho}_2)}{(\hat{\rho}_3 - \hat{\rho}_1)(\hat{\rho}_3 - \hat{\rho}_2)}. \tag{5.183}$$

As an example of interpolation, $\widetilde{u}^m_{j_a+1}$ is shown in Fig. 5.5 which is obtained for $\Delta\rho_a = 10^{-5}$, $\Delta\rho_b = 10^{-4}$, $\hat{\rho}_1 = 10^{-4}$, $\hat{u}_1 = 0.9$, $\hat{u}_2 = 0.3$, and $\hat{u}_3 = 0.2$.

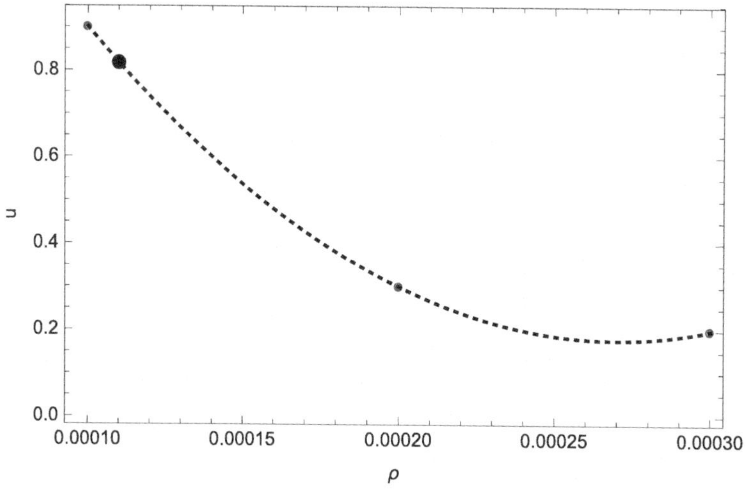

Figure 5.5 An example of interpolation of $\widetilde{u}^m_{j_a+1}$ in the ρ–u plane. Small points represent $(\hat{\rho}_j, \hat{u}_j)$ $(j = 1, 2, 3)$ in order from the left. A large point represents a virtual point $(\hat{\rho}_x, \widetilde{u}^m_{j_a+1})$ to be interpolated.

- In the case of $j = j_a$,

Applying the above interpolation method and substituting Eq. (5.182) into Eq. (5.180) yields the following:

$$
\begin{aligned}
-2\frac{\mathcal{B}(\rho_j, u_j^n)}{\Delta\beta}u_j^n &= \frac{\rho_j}{(\Delta\rho_a)^2}u_{j-1}^{n+1/2} - 2\left[\frac{\rho_j}{(\Delta\rho_a)^2} + \frac{\mathcal{B}(\rho_j, u_j^n)}{\Delta\beta}\right]u_j^{n+1/2} \\
&\quad + \frac{\rho_j}{(\Delta\rho_a)^2}\tilde{u}_{j+1}^{n+1/2} \\
&= \frac{\rho_j}{(\Delta\rho_a)^2}u_{j-1}^{n+1/2} - 2\left[\frac{\rho_j}{(\Delta\rho_a)^2} + \frac{\mathcal{B}(\rho_j, u_j^n)}{\Delta\beta}\right]u_j^{n+1/2} \\
&\quad + \frac{\rho_j}{(\Delta\rho_a)^2}\left[\hat{c}_{u1}u_j^{n+1/2} + \hat{c}_{u2}u_{j+1}^{n+1/2} + \hat{c}_{u3}u_{j+2}^{n+1/2}\right] \\
&= \frac{\rho_j}{(\Delta\rho_a)^2}u_{j-1}^{n+1/2} - \left[\frac{\rho_j}{(\Delta\rho_a)^2}(2 - \hat{c}_{u1}) + 2\frac{\mathcal{B}(\rho_j, u_j^n)}{\Delta\beta}\right]u_j^{n+1/2} \\
&\quad + \frac{\rho_j}{(\Delta\rho_a)^2}\hat{c}_{u2}u_{j+1}^{n+1/2} + \frac{\rho_j}{(\Delta\rho_a)^2}\hat{c}_{u3}u_{j+2}^{n+1/2}.
\end{aligned} \tag{5.184}
$$

5.10.1.3 Corrector

Corrector Eq. (5.134) can be classified as follows:

- In the case of $1 \le j \le j_a - 1$,

$$
\begin{aligned}
&\frac{\rho_j}{(\Delta\rho_a)^2}u_{j-1}^{n+1} - 2\left[\frac{\rho_j}{(\Delta\rho_a)^2} + \frac{\mathcal{B}(\rho_j, u_j^{n+1/2})}{\Delta\beta}\right]u_j^{n+1} + \frac{\rho_j}{(\Delta\rho_a)^2}u_{j+1}^{n+1} \\
&= -\frac{\rho_j}{(\Delta\rho_a)^2}u_{j+1}^n + 2\left[\frac{\rho_j}{(\Delta\rho_a)^2} - \frac{\mathcal{B}(\rho_j, u_j^{n+1/2})}{\Delta\beta}\right]u_j^n \\
&\quad - \frac{\rho_j}{(\Delta\rho_a)^2}u_{j-1}^n
\end{aligned} \tag{5.185}
$$

- In the case of $j = j_a$,

$$\frac{\rho_j}{(\Delta\rho_a)^2}u_{j-1}^{n+1} - 2\left[\frac{\rho_j}{(\Delta\rho_a)^2} + \frac{\mathcal{B}(\rho_j, u_j^{n+1/2})}{\Delta\beta}\right]u_j^{n+1} + \frac{\rho_j}{(\Delta\rho_a)^2}\widetilde{u}_{j+1}^{n+1}$$
$$= \frac{\rho_j}{(\Delta\rho_a)^2}u_{j-1}^{n+1} - 2\left[\frac{\rho_j}{(\Delta\rho_a)^2} + \frac{\mathcal{B}(\rho_j, u_j^{n+1/2})}{\Delta\beta}\right]u_j^{n+1}$$
$$+\frac{\rho_j}{(\Delta\rho_a)^2}(\hat{c}_{u1}u_j^{n+1} + \hat{c}_{u2}u_{j+1}^{n+1} + \hat{c}_{u3}u_{j+2}^{n+1})$$
$$= \frac{\rho_j}{(\Delta\rho_a)^2}u_{j-1}^{n+1} - \left[\frac{\rho_j}{(\Delta\rho_a)^2}(2 - \hat{c}_{u1}) + 2\frac{\mathcal{B}(\rho_j, u_j^{n+1/2})}{\Delta\beta}\right]u_j^{n+1}$$
$$+\frac{\rho_j}{(\Delta\rho_a)^2}\hat{c}_{u2}u_{j+1}^{n+1} + \frac{\rho_j}{(\Delta\rho_a)^2}\hat{c}_{u3}u_{j+2}^{n+1}$$
$$= -\frac{\rho_j}{(\Delta\rho_a)^2}\widetilde{u}_{j+1}^{n} + 2\left[\frac{\rho_j}{(\Delta\rho_a)^2} - \frac{\mathcal{B}(\rho_j, u_j^{n+1/2})}{\Delta\beta}\right]u_j^{n} - \frac{\rho_j}{(\Delta\rho_a)^2}u_{j-1}^{n} \tag{5.186}$$

where,

$$\widetilde{u}_{j+1}^{n} = u(\rho_j + \Delta\rho_a) = \hat{c}_{u1}u_j^n + \hat{c}_{u2}u_{j+1}^n + \hat{c}_{u3}u_{j+2}^n. \tag{5.187}$$

- In the case of $j_a + 1 \le j \le J - 1$,

$$\frac{\rho_j}{(\Delta\rho_b)^2}u_{j-1}^{n+1} - 2\left[\frac{\rho_j}{(\Delta\rho_b)^2} + \frac{\mathcal{B}(\rho_j, u_j^{n+1/2})}{\Delta\beta}\right]u_j^{n+1} + \frac{\rho_j}{(\Delta\rho_b)^2}u_{j+1}^{n+1}$$
$$= -\frac{\rho_j}{(\Delta\rho_b)^2}u_{j+1}^{n} + 2\left[\frac{\rho_j}{(\Delta\rho_b)^2} - \frac{\mathcal{B}(\rho_j, u_j^{n+1/2})}{\Delta\beta}\right]u_j^{n} - \frac{\rho_j}{(\Delta\rho_b)^2}u_{j-1}^{n} \tag{5.188}$$

- In the case of $j = J$,

$$u_{J-2}^{n+1} - 2u_{J-1}^{n+1} + u_J^{n+1} = \Delta^2 u^0. \tag{5.189}$$

5.10.1.4 Simultaneous linear equations for u_j ($1 \le j \le J$)

The excess internal energy u_j ($1 \le j \le J$) for Predictor and Collector described above are obtained as solutions of the following simultaneous linear equations. Here, for the sake of simplicity, the case of $j_a = 5$ and $J = 9$ will be explained. But the obtained results are easy to extend to realistic cases where J and j_a are sufficiently larger than 9 but j_a is sufficiently smaller than J.

5.10.1.5 LU decomposition of matrix A

Arbitrary square matrix A can be decomposed into $A = \hat{L}U$ using lower triangular matrix $\hat{L}$ and upper triangular matrix U. Here, the coefficient matrix for $J = 9$ and $j_a = 5$ is decomposed into $A = \hat{L}U$. In this case, A, $\hat{L}$, and U can be written as follows:

$$A = \begin{bmatrix} b_1 & c_1 & 0 & 0 & 0 & 0 & 0 & 0 & 0 \\ a_2 & b_2 & c_2 & 0 & 0 & 0 & 0 & 0 & 0 \\ 0 & a_3 & b_3 & c_3 & 0 & 0 & 0 & 0 & 0 \\ 0 & 0 & a_4 & b_4 & c_4 & 0 & 0 & 0 & 0 \\ 0 & 0 & 0 & a_5 & b_5 & c_5 & d_5 & 0 & 0 \\ 0 & 0 & 0 & 0 & a_6 & b_6 & c_6 & 0 & 0 \\ 0 & 0 & 0 & 0 & 0 & a_7 & b_7 & c_7 & 0 \\ 0 & 0 & 0 & 0 & 0 & 0 & a_8 & b_8 & c_8 \\ 0 & 0 & 0 & 0 & 0 & 0 & 1 & -2 & 1 \end{bmatrix},$$

$$\hat{L} = \begin{bmatrix} L_{11} & 0 & 0 & 0 & 0 & 0 & 0 & 0 & 0 \\ L_{21} & L_{22} & 0 & 0 & 0 & 0 & 0 & 0 & 0 \\ L_{31} & L_{32} & L_{33} & 0 & 0 & 0 & 0 & 0 & 0 \\ L_{41} & L_{42} & L_{43} & L_{44} & 0 & 0 & 0 & 0 & 0 \\ L_{51} & L_{52} & L_{53} & L_{54} & L_{55} & 0 & 0 & 0 & 0 \\ L_{61} & L_{62} & L_{63} & L_{64} & L_{65} & L_{66} & 0 & 0 & 0 \\ L_{71} & L_{72} & L_{73} & L_{74} & L_{75} & L_{76} & L_{77} & 0 & 0 \\ L_{81} & L_{82} & L_{83} & L_{84} & L_{85} & L_{86} & L_{87} & L_{88} & 0 \\ L_{91} & L_{92} & L_{93} & L_{94} & L_{95} & L_{96} & L_{97} & L_{98} & L_{99} \end{bmatrix},$$

$$U = \begin{bmatrix} 1 & U_{12} & U_{13} & U_{14} & U_{15} & U_{16} & U_{17} & U_{18} & U_{19} \\ 0 & 1 & U_{23} & U_{24} & U_{25} & U_{26} & U_{27} & U_{28} & U_{29} \\ 0 & 0 & 1 & U_{34} & U_{35} & U_{36} & U_{37} & U_{38} & U_{39} \\ 0 & 0 & 0 & 1 & U_{45} & U_{46} & U_{47} & U_{48} & U_{49} \\ 0 & 0 & 0 & 0 & 1 & U_{56} & U_{57} & U_{58} & U_{59} \\ 0 & 0 & 0 & 0 & 0 & 1 & U_{67} & U_{68} & U_{69} \\ 0 & 0 & 0 & 0 & 0 & 0 & 1 & U_{78} & U_{79} \\ 0 & 0 & 0 & 0 & 0 & 0 & 0 & 1 & U_{89} \\ 0 & 0 & 0 & 0 & 0 & 0 & 0 & 0 & 1 \end{bmatrix}.$$

From the relation $A_{ij} = \sum_{k=1}^{J} L_{ik}U_{kj}$, we can get the following:

$$\hat{L} = \begin{bmatrix} L_{11} & 0 & 0 & 0 & 0 & 0 & 0 & 0 & 0 \\ L_{21} & L_{22} & 0 & 0 & 0 & 0 & 0 & 0 & 0 \\ 0 & L_{32} & L_{33} & 0 & 0 & 0 & 0 & 0 & 0 \\ 0 & 0 & L_{43} & L_{44} & 0 & 0 & 0 & 0 & 0 \\ 0 & 0 & 0 & L_{54} & L_{55} & 0 & 0 & 0 & 0 \\ 0 & 0 & 0 & 0 & L_{65} & L_{66} & 0 & 0 & 0 \\ 0 & 0 & 0 & 0 & 0 & L_{76} & L_{77} & 0 & 0 \\ 0 & 0 & 0 & 0 & 0 & 0 & L_{87} & L_{88} & 0 \\ 0 & 0 & 0 & 0 & 0 & 0 & 1 & L_{98} & L_{99} \end{bmatrix}, \tag{5.190}$$

$$U = \begin{bmatrix} 1 & U_{12} & 0 & 0 & 0 & 0 & 0 & 0 & 0 \\ 0 & 1 & U_{23} & 0 & 0 & 0 & 0 & 0 & 0 \\ 0 & 0 & 1 & U_{34} & 0 & 0 & 0 & 0 & 0 \\ 0 & 0 & 0 & 1 & U_{45} & 0 & 0 & 0 & 0 \\ 0 & 0 & 0 & 0 & 1 & U_{56} & U_{57} & 0 & 0 \\ 0 & 0 & 0 & 0 & 0 & 1 & U_{67} & 0 & 0 \\ 0 & 0 & 0 & 0 & 0 & 0 & 1 & U_{78} & 0 \\ 0 & 0 & 0 & 0 & 0 & 0 & 0 & 1 & U_{89} \\ 0 & 0 & 0 & 0 & 0 & 0 & 0 & 0 & 1 \end{bmatrix}, \tag{5.191}$$

where,

$$\begin{cases} L_{11} = b_1, \quad L_{i,i-1} = a_i \ (2 \le i \le J-1), \\ L_{ii} = b_i - U_{i-1,i}a_i \quad (2 \le i \le J-1), \\ L_{J,J-2} = 1, \ L_{J,J-1} = -2 - U_{J-2,J-1}, \\ L_{JJ} = 1 + c_{J-1}\dfrac{2 + U_{J-2,J-1}}{b_{J-1} - U_{J-2,J-1}a_{J-1}}. \end{cases} \tag{5.192}$$

$$\begin{cases} U_{ii} = 1\,(1 \le i \le J), \ U_{12} = \dfrac{c_1}{b_1}, \\ U_{i,i+1} = \dfrac{c_i}{b_i - U_{i-1,i}a_i} \ (2 \le i \le j_a \ \text{or} \ j_a + 2 \le i \le J-1), \\ U_{j_a,j_a+2} = \dfrac{d_{j_a}}{b_{j_a} - U_{j_a-1,j_a}a_{j_a}}, \\ U_{j_a+1,j_a+2} = \dfrac{c_{j_a+1} - U_{j_a,j_a+2}a_{j_a+1}}{b_{j_a+1} - U_{j_a,j_a+1}a_{j_a+1}}. \end{cases} \tag{5.193}$$

5.10.1.6 Solutions of $A\mathbf{u} = \mathbf{r}$

If we set $\mathbf{u}^{\mathrm{t}} = (u_1, u_2, \cdots, u_J)$ and $\mathbf{r}^{\mathrm{t}} = (r_1, r_2, \cdots, r_J)$, the equation to be solved can be written as $A\mathbf{u} = \mathbf{r}$, where $\mathbf{u}^{\mathrm{t}}$ and $\mathbf{r}^{\mathrm{t}}$ are the transposes of $\mathbf{u}$ and $\mathbf{r}$, respectively. If we set $U\mathbf{u} = \mathbf{h}$, we can write $A\mathbf{u} = \hat{L}U\mathbf{u} = \hat{L}\mathbf{h} = \mathbf{r}$, and we get the following:

$$\hat{L}\mathbf{h} = \begin{bmatrix} L_{11}h_1 \\ L_{21}h_1 + L_{22}h_2 \\ L_{32}h_2 + L_{33}h_3 \\ L_{43}h_3 + L_{44}h_4 \\ L_{54}h_4 + L_{55}h_5 \\ L_{65}h_5 + L_{66}h_6 \\ L_{76}h_6 + L_{77}h_7 \\ L_{87}h_7 + L_{88}h_8 \\ h_7 + L_{98}h_8 + L_{99}h_9 \end{bmatrix} = \begin{bmatrix} r_1 \\ r_2 \\ r_3 \\ r_4 \\ r_5 \\ r_6 \\ r_7 \\ r_8 \\ r_9 \end{bmatrix},$$

where,

$$\mathbf{h}^{\mathrm{t}} = [h_1, h_2, h_3, \cdots, h_9].$$

Therefore,

$$h_1 = \frac{r_1}{L_{11}}, \quad h_j = \frac{r_j - L_{j,j-1}h_{j-1}}{L_{jj}} \quad (2 \le j \le J-1),$$
$$h_J = \frac{r_J - h_{J-2} - L_{J,J-1}h_{J-1}}{L_{JJ}}. \tag{5.194}$$

And, from $U\mathbf{u} = \mathbf{h}$,

$$\begin{bmatrix} u_1 + U_{12}u_2 \\ u_2 + U_{23}u_3 \\ u_3 + U_{34}u_4 \\ u_4 + U_{45}u_5 \\ u_5 + U_{56}u_6 + U_{57}u_7 \\ u_6 + U_{67}u_7 \\ u_7 + U_{78}u_8 \\ u_8 + U_{89}u_9 \\ u_9 \end{bmatrix} = \begin{bmatrix} h_1 \\ h_2 \\ h_3 \\ h_4 \\ h_5 \\ h_6 \\ h_7 \\ h_8 \\ h_9 \end{bmatrix}.$$

Therefore,

$$u_J = h_J,$$
$$u_j = h_j - U_{j,j+1}u_{j+1} \quad (j = J-1, J-2, \cdots, j_a+1),$$
$$u_{j_a} = h_{j_a} - U_{j_a,j_a+1}u_{j_a+1} - U_{j_a,j_a+2}u_{j_a+2},$$
$$u_j = h_j - U_{j,j+1}u_{j+1} \quad (j = j_a-1, j_a-2, \cdots, 1). \tag{5.195}$$

Here, in order to simplify the expressions, new variables p_j $(1 \leq j \leq J-1)$ and q are introduced as follows:

$$
\begin{cases}
p_1 = U_{12} = \dfrac{c_1}{b_1}, \\
p_j = U_{j,j+1} = \dfrac{c_j}{b_j - a_j p_{j-1}} \quad (j = 2, 3, \cdots, j_a), \\
q = U_{j_a, j_a+2} = \dfrac{d_{j_a}}{b_{j_a} - a_{j_a} p_{j_a-1}}, \\
p_{j_a+1} = U_{j_a+1, j_a+2} = \dfrac{c_{j_a+1} - a_{j_a+1} q}{b_{j_a+1} - a_{j_a+1} p_{j_a}}, \\
p_j = U_{j,j+1} = \dfrac{c_j}{b_j - a_j p_{j-1}} \quad (j = j_a + 2, j_a + 3, \cdots, J-1),
\end{cases}
$$

then,

$$
\begin{cases}
h_1 = \dfrac{r_1}{b_1}, \\
h_j = \dfrac{r_j - h_{j-1} a_j}{b_j - a_j p_{j-1}} \quad (j = 2, 3, \cdots, J-1), \\
h_J = \dfrac{r_J - h_{J-2} + h_{J-1}(2 + p_{J-2})}{1 + (2 + p_{J-2}) p_{J-1}}.
\end{cases}
\tag{5.196}
$$

The following equations are finally obtained:

$$
\begin{cases}
u_J = h_J, \\
u_j = h_j - p_j u_{j+1}, \quad (j = J-1, J-2, \cdots, j_a + 1), \\
u_{j_a} = h_{j_a} - p_{j_a} u_{j_a+1} - q u_{j_a+2}, \\
u_j = h_j - p_j u_{j+1}, \quad (j = j_a - 1, j_a - 2, \cdots, 1).
\end{cases}
\tag{5.197}
$$

Here, for Predictor, a_j $(2 \leq j \leq J-1)$, b_j, c_j $(1 \leq j \leq J-1)$, r_j $(1 \leq j \leq J)$, and d_{j_a} are given by

- In the case of $(1 \leq j \leq j_a - 1)$,

$$
\begin{aligned}
& a_j = c_j = \frac{\rho_j}{(\Delta\rho_a)^2}, \quad b_j = -2\left[\frac{\rho_j}{(\Delta\rho_a)^2} + \frac{\mathcal{B}(\rho_j, u_j^n)}{\Delta\beta}\right], \\
& r_j = -2\frac{\mathcal{B}(\rho_j, u_j^n)}{\Delta\beta} u_j^n.
\end{aligned}
\tag{5.198}
$$

Here, a_1 is a coefficient that is not actually used.

- In the case of $j = j_a$,

$$a_j = \frac{\rho_j}{(\Delta\rho_a)^2},\ b_j = -\left[\frac{\rho_j}{(\Delta\rho_a)^2}(2-\hat{c}_{u1}) + 2\frac{\mathcal{B}(\rho_j, u_j^n)}{\Delta\beta}\right],$$
$$c_j = \frac{\rho_j}{(\Delta\rho_a)^2}\hat{c}_{u2},\ d_j = \frac{\rho_j}{(\Delta\rho_a)^2}\hat{c}_{u3},\ r_j = -2\frac{\mathcal{B}(\rho_j, u_j^n)}{\Delta\beta}u_j^n. \quad (5.199)$$

- In the case of $(j_a + 1 \le j \le J - 1)$,

$$a_j = c_j = \frac{\rho_j}{(\Delta\rho_b)^2},\ b_j = -2\left[\frac{\rho_j}{(\Delta\rho_b)^2} + \frac{\mathcal{B}(\rho_j, u_j^n)}{\Delta\beta}\right],$$
$$r_j = -2\frac{\mathcal{B}(\rho_j, u_j^n)}{\Delta\beta}u_j^n. \quad (5.200)$$

- In the case of $j = J$,

$$r_j = \Delta^2 u^0. \quad (5.201)$$

For the Collector, these coefficients are given by the following equations:

- In the case of $(1 \le j \le j_a - 1)$,

$$a_j = c_j = \frac{\rho_j}{(\Delta\rho_a)^2},\ b_j = -2\left[\frac{\rho_j}{(\Delta\rho_a)^2} + \frac{\mathcal{B}(\rho_j, u_j^{n+1/2})}{\Delta\beta}\right],$$
$$r_j = -\frac{\rho_j}{(\Delta\rho_a)^2}u_{j+1}^n + 2\left[\frac{\rho_j}{(\Delta\rho_a)^2} - \frac{\mathcal{B}(\rho_j, u_j^{n+1/2})}{\Delta\beta}\right]u_j^n$$
$$-\frac{\rho_j}{(\Delta\rho_a)^2}u_{j-1}^n. \quad (5.202)$$

Here, a_1 is a coefficient that is not actually used.

- In the case of $j = j_a$,

$$a_j = \frac{\rho_j}{(\Delta\rho_a)^2},\ b_j = -\left[\frac{\rho_j}{(\Delta\rho_a)^2}(2-\hat{c}_{u1}) + 2\frac{\mathcal{B}(\rho_j, u_j^{n+1/2})}{\Delta\beta}\right],$$
$$c_j = \frac{\rho_j}{(\Delta\rho_a)^2}\hat{c}_{u2},\ d_j = \frac{\rho_j}{(\Delta\rho_a)^2}\hat{c}_{u3},\ r_j = -\frac{\rho_j}{(\Delta\rho_a)^2}\widetilde{u}_{j+1}^n$$
$$+2\left[\frac{\rho_j}{(\Delta\rho_a)^2} - \frac{\mathcal{B}(\rho_j, u_j^{n+1/2})}{\Delta\beta}\right]u_j^n - \frac{\rho_j}{(\Delta\rho_a)^2}u_{j-1}^n, \quad (5.203)$$

where,

$$\widetilde{u}_{j+1}^n = \hat{c}_{u1}u_j^n + \hat{c}_{u2}u_{j+1}^n + \hat{c}_{u3}u_{j+2}^n.$$

- In the case of $(j_a + 1 \le j \le J - 1)$,

$$a_j = c_j = \frac{\rho_j}{(\Delta\rho_b)^2},\ b_j = -2\left[\frac{\rho_j}{(\Delta\rho_b)^2} + \frac{\mathcal{B}(\rho_j, u_j^{n+1/2})}{\Delta\beta}\right],$$

$$r_j = -\frac{\rho_j}{(\Delta\rho_b)^2}u_{j+1}^n + 2\left[\frac{\rho_j}{(\Delta\rho_b)^2} - \frac{\mathcal{B}(\rho_j, u_j^{n+1/2})}{\Delta\beta}\right]u_j^n$$
$$-\frac{\rho_j}{(\Delta\rho_b)^2}u_{j-1}^n. \tag{5.204}$$

- In the case of $j = J$,

$$r_j = \Delta^2 u^0. \tag{5.205}$$

5.10.2 In the Case of $T < T_c$

In this case, $\mathcal{B}(\rho, u)$ has positive values in the region of $1 < j \le L$ and $R \le j \le J$, but it takes negative values in the region of $L < j < R$. This section describes the case of $1 \le j \le L$, assuming that j_a and L are sufficiently larger than 1 but j_a is sufficiently smaller than L.

5.10.2.1 Predictor for $1 \le j \le L$

- In the case of $1 \le j \le j_a - 1$,

$$\frac{\rho_j}{(\Delta\rho_a)^2}u_{j-1}^{n+1/2} - 2\left[\frac{\rho_j}{(\Delta\rho_a)^2} + \frac{\mathcal{B}(\rho_j, u_j^n)}{\Delta\beta}\right]u_j^{n+1/2} + \frac{\rho_j}{(\Delta\rho_a)^2}u_{j+1}^{n+1/2}$$
$$= -2\frac{\mathcal{B}(\rho_j, u_j^n)}{\Delta\beta}u_j^n. \tag{5.206}$$

- In the case of $j = j_a$,

$$\frac{\rho_j}{(\Delta\rho_a)^2}u_{j-1}^{n+1/2} - \left[\frac{\rho_j}{(\Delta\rho_a)^2}(2 - \hat{c}_{u1}) + 2\frac{\mathcal{B}(\rho_j, u_j^n)}{\Delta\beta}\right]u_j^{n+1/2}$$
$$+\frac{\rho_j}{(\Delta\rho_a)^2}\hat{c}_{u2}u_{j+1}^{n+1/2} + \frac{\rho_j}{(\Delta\rho_a)^2}\hat{c}_{u3}u_{j+2}^{n+1/2}$$
$$= -2\frac{\mathcal{B}(\rho_j, u_j^n)}{\Delta\beta}u_j^n. \tag{5.207}$$

- In the case of $j_a + 1 \le j \le L - 1$,

$$\frac{\rho_j}{(\Delta\rho_b)^2}u_{j-1}^{n+1/2} - 2\left[\frac{\rho_j}{(\Delta\rho_b)^2} + \frac{\mathcal{B}(\rho_j, u_j^n)}{\Delta\beta}\right]u_j^{n+1/2} + \frac{\rho_j}{(\Delta\rho_b)^2}u_{j+1}^{n+1/2}$$
$$= -2\frac{\mathcal{B}(\rho_j, u_j^n)}{\Delta\beta}u_j^n. \tag{5.208}$$

• In the case of $j = L$,

$$\frac{\rho_L}{(\Delta\rho_b)^2}u_{L-1}^{n+1/2} - 2\left[\frac{\rho_L}{(\Delta\rho_b)^2} + \frac{\mathcal{B}(\rho_L, u_L^n)}{\Delta\beta}\right]u_L^{n+1/2}$$
$$= -2\frac{\mathcal{B}(\rho_L, u_L^n)}{\Delta\beta}u_L^n - \frac{\rho_L}{(\Delta\rho_b)^2}u_S(\rho_{L+1}). \qquad (5.209)$$

5.10.2.2 Corrector for $1 \le j \le L$

• In the case of $1 \le j \le j_a - 1$,

$$\frac{\rho_j}{(\Delta\rho_a)^2}u_{j-1}^{n+1} - 2\left[\frac{\rho_j}{(\Delta\rho_a)^2} + \frac{\mathcal{B}(\rho_j, u_j^{n+1/2})}{\Delta\beta}\right]u_j^{n+1} + \frac{\rho_j}{(\Delta\rho_a)^2}u_{j+1}^{n+1}$$
$$= -\frac{\rho_j}{(\Delta\rho_a)^2}u_{j+1}^n + 2\left[\frac{\rho_j}{(\Delta\rho_a)^2} - \frac{\mathcal{B}(\rho_j, u_j^{n+1/2})}{\Delta\beta}\right]u_j^n$$
$$-\frac{\rho_j}{(\Delta\rho_a)^2}u_{j-1}^n. \qquad (5.210)$$

• In the case of $j = j_a$,

$$\frac{\rho_j}{(\Delta\rho_a)^2}u_{j-1}^{n+1} - \left[\frac{\rho_j}{(\Delta\rho_a)^2}(2 - \hat{c}_{u1}) + 2\frac{\mathcal{B}(\rho_j, u_j^{n+1/2})}{\Delta\beta}\right]u_j^{n+1}$$
$$+\frac{\rho_j}{(\Delta\rho_a)^2}\hat{c}_{u2}u_{j+1}^{n+1} + \frac{\rho_j}{(\Delta\rho_a)^2}\hat{c}_{u3}u_{j+2}^{n+1}$$
$$= -\frac{\rho_j}{(\Delta\rho_a)^2}\widetilde{u}_{j+1}^n + 2\left[\frac{\rho_j}{(\Delta\rho_a)^2} - \frac{\mathcal{B}(\rho_j, u_j^{n+1/2})}{\Delta\beta}\right]u_j^n$$
$$-\frac{\rho_j}{(\Delta\rho_a)^2}u_{j-1}^n \qquad (5.211)$$

$$\widetilde{u}_{j+1}^n = u(\rho_j + \Delta\rho_a) = \hat{c}_{u1}u_j^n + \hat{c}_{u2}u_{j+1}^n + \hat{c}_{u3}u_{j+2}^n. \qquad (5.212)$$

• In the case of $j_a + 1 \le j \le L - 1$,

$$\frac{\rho_j}{(\Delta\rho_b)^2}u_{j-1}^{n+1} - 2\left[\frac{\rho_j}{(\Delta\rho_b)^2} + \frac{\mathcal{B}(\rho_j, u_j^{n+1/2})}{\Delta\beta}\right]u_j^{n+1} + \frac{\rho_j}{(\Delta\rho_b)^2}u_{j+1}^{n+1}$$
$$= -\frac{\rho_j}{(\Delta\rho_b)^2}u_{j+1}^n + 2\left[\frac{\rho_j}{(\Delta\rho_b)^2} - \frac{\mathcal{B}(\rho_j, u_j^{n+1/2})}{\Delta\beta}\right]u_j^n$$
$$-\frac{\rho_j}{(\Delta\rho_b)^2}u_{j-1}^n. \qquad (5.213)$$

• In the case of $j = L$,

$$\frac{\rho_j}{(\Delta\rho_b)^2}u_{j-1}^{n+1} - 2\left[\frac{\rho_j}{(\Delta\rho_b)^2} + \frac{\mathcal{B}(\rho_j, u_j^{n+1/2})}{\Delta\beta}\right]u_j^{n+1}$$

$$= -2\frac{\rho_j}{(\Delta\rho_b)^2}u_S(\rho_{L+1}) + 2\left[\frac{\rho_j}{(\Delta\rho_b)^2} - \frac{\mathcal{B}(\rho_j, u_j^{n+1/2})}{\Delta\beta}\right]u_j^n$$

$$-\frac{\rho_j}{(\Delta\rho_b)^2}u_{j-1}^n. \tag{5.214}$$

5.10.2.3 Simultaneous linear equations for u_j $(1 \le j \le L)$

Both u_j $(1 \le j \le L)$ for Predictor and Collector described above are obtained as solutions of the following simultaneous linear equations. Here, for the sake of simplicity, the case of $j_a = 5$ and $L = 9$ will be described. It is easy to extend the obtained results for any L and j_a, assuming that j_a and L are sufficiently larger than 5 but j_a is sufficiently smaller than L.

5.10.2.4 LU decomposition of matrix A

Arbitrary square matrix A can be decomposed into $A = \hat{L}U$ using lower triangular matrix $\hat{L}$ and upper triangular matrix U. Here, the coefficient matrix for $L = 9$ and $j_a = 5$ is decomposed into $A = \hat{L}U$. In this case, A, $\hat{L}$, and U can be written as follows:

$$A = \begin{bmatrix} b_1 & c_1 & 0 & 0 & 0 & 0 & 0 & 0 & 0 \\ a_2 & b_2 & c_2 & 0 & 0 & 0 & 0 & 0 & 0 \\ 0 & a_3 & b_3 & c_3 & 0 & 0 & 0 & 0 & 0 \\ 0 & 0 & a_4 & b_4 & c_4 & 0 & 0 & 0 & 0 \\ 0 & 0 & 0 & a_5 & b_5 & c_5 & d_5 & 0 & 0 \\ 0 & 0 & 0 & 0 & a_6 & b_6 & c_6 & 0 & 0 \\ 0 & 0 & 0 & 0 & 0 & a_7 & b_7 & c_7 & 0 \\ 0 & 0 & 0 & 0 & 0 & 0 & a_8 & b_8 & c_8 \\ 0 & 0 & 0 & 0 & 0 & 0 & 0 & a_9 & b_9 \end{bmatrix},$$

$$\hat{L} = \begin{bmatrix} L_{11} & 0 & 0 & 0 & 0 & 0 & 0 & 0 & 0 \\ L_{21} & L_{22} & 0 & 0 & 0 & 0 & 0 & 0 & 0 \\ L_{31} & L_{32} & L_{33} & 0 & 0 & 0 & 0 & 0 & 0 \\ L_{41} & L_{42} & L_{43} & L_{44} & 0 & 0 & 0 & 0 & 0 \\ L_{51} & L_{52} & L_{53} & L_{54} & L_{55} & 0 & 0 & 0 & 0 \\ L_{61} & L_{62} & L_{63} & L_{64} & L_{65} & L_{66} & 0 & 0 & 0 \\ L_{71} & L_{72} & L_{73} & L_{74} & L_{75} & L_{76} & L_{77} & 0 & 0 \\ L_{81} & L_{82} & L_{83} & L_{84} & L_{85} & L_{86} & L_{87} & L_{88} & 0 \\ L_{91} & L_{92} & L_{93} & L_{94} & L_{95} & L_{96} & L_{97} & L_{98} & L_{99} \end{bmatrix},$$

$$U = \begin{bmatrix} 1 & U_{12} & U_{13} & U_{14} & U_{15} & U_{16} & U_{17} & U_{18} & U_{19} \\ 0 & 1 & U_{23} & U_{24} & U_{25} & U_{26} & U_{27} & U_{28} & U_{29} \\ 0 & 0 & 1 & U_{34} & U_{35} & U_{36} & U_{37} & U_{38} & U_{39} \\ 0 & 0 & 0 & 1 & U_{45} & U_{46} & U_{47} & U_{48} & U_{49} \\ 0 & 0 & 0 & 0 & 1 & U_{56} & U_{57} & U_{58} & U_{59} \\ 0 & 0 & 0 & 0 & 0 & 1 & U_{67} & U_{68} & U_{69} \\ 0 & 0 & 0 & 0 & 0 & 0 & 1 & U_{78} & U_{79} \\ 0 & 0 & 0 & 0 & 0 & 0 & 0 & 1 & U_{89} \\ 0 & 0 & 0 & 0 & 0 & 0 & 0 & 0 & 1 \end{bmatrix}.$$

Because $A_{ij} = \sum_{k=1}^{L} L_{ik}U_{kj}$, we get the following:

$$\hat{L} = \begin{bmatrix} L_{11} & 0 & 0 & 0 & 0 & 0 & 0 & 0 & 0 \\ L_{21} & L_{22} & 0 & 0 & 0 & 0 & 0 & 0 & 0 \\ 0 & L_{32} & L_{33} & 0 & 0 & 0 & 0 & 0 & 0 \\ 0 & 0 & L_{43} & L_{44} & 0 & 0 & 0 & 0 & 0 \\ 0 & 0 & 0 & L_{54} & L_{55} & 0 & 0 & 0 & 0 \\ 0 & 0 & 0 & 0 & L_{65} & L_{66} & 0 & 0 & 0 \\ 0 & 0 & 0 & 0 & 0 & L_{76} & L_{77} & 0 & 0 \\ 0 & 0 & 0 & 0 & 0 & 0 & L_{87} & L_{88} & 0 \\ 0 & 0 & 0 & 0 & 0 & 0 & 0 & L_{98} & L_{99} \end{bmatrix}, \tag{5.215}$$

$$U = \begin{bmatrix} 1 & U_{12} & 0 & 0 & 0 & 0 & 0 & 0 & 0 \\ 0 & 1 & U_{23} & 0 & 0 & 0 & 0 & 0 & 0 \\ 0 & 0 & 1 & U_{34} & 0 & 0 & 0 & 0 & 0 \\ 0 & 0 & 0 & 1 & U_{45} & 0 & 0 & 0 & 0 \\ 0 & 0 & 0 & 0 & 1 & U_{56} & U_{57} & 0 & 0 \\ 0 & 0 & 0 & 0 & 0 & 1 & U_{67} & 0 & 0 \\ 0 & 0 & 0 & 0 & 0 & 0 & 1 & U_{78} & 0 \\ 0 & 0 & 0 & 0 & 0 & 0 & 0 & 1 & U_{89} \\ 0 & 0 & 0 & 0 & 0 & 0 & 0 & 0 & 1 \end{bmatrix}, \tag{5.216}$$

where,

$$\begin{cases} L_{11} = b_1, \quad L_{i,i-1} = a_i \ (2 \leq i \leq L), \\ L_{ii} = b_i - U_{i-1,i} a_i \quad (2 \leq i \leq L). \end{cases} \tag{5.217}$$

$$\begin{cases} U_{12} = \dfrac{c_1}{b_1}, \\ U_{i,i+1} = \dfrac{c_i}{b_i - U_{i-1,i} a_i} \ (2 \leq i \leq j_a), \\ U_{j_a,\, j_a+2} = \dfrac{d_{j_a}}{b_{j_a} - U_{j_a-1,\, j_a} a_{j_a}}, \\ U_{j_a+1,\, j_a+2} = \dfrac{c_{j_a+1} - U_{j_a,\, j_a+2} a_{j_a+1}}{b_{j_a+1} - U_{j_a,\, j_a+1} a_{j_a+1}}, \\ U_{i,i+1} = \dfrac{c_i}{b_i - U_{i-1,i} a_i} \ (j_a + 2 \leq i \leq L-1). \end{cases} \tag{5.218}$$

5.10.2.5 Solutions of $A\mathbf{u} = \mathbf{r}$

The equations to be solved can be written as $A\mathbf{u} = \mathbf{r}$. Here, we set $U\mathbf{u} = \mathbf{h}$ to get $A\mathbf{u} = \hat{L}U\mathbf{u} = \hat{L}\mathbf{h} = \mathbf{r}$. This leads to following:

$$\hat{L}\mathbf{h} = \begin{bmatrix} L_{11}h_1 \\ L_{21}h_1 + L_{22}h_2 \\ L_{32}h_2 + L_{33}h_3 \\ L_{43}h_3 + L_{44}h_4 \\ L_{54}h_4 + L_{55}h_5 \\ L_{65}h_5 + L_{66}h_6 \\ L_{76}h_6 + L_{77}h_7 \\ L_{87}h_7 + L_{88}h_8 \\ L_{98}h_8 + L_{99}h_9 \end{bmatrix} = \begin{bmatrix} r_1 \\ r_2 \\ r_3 \\ r_4 \\ r_5 \\ r_6 \\ r_7 \\ r_8 \\ r_9 \end{bmatrix},$$

where,

$$\hat{L}\mathbf{h} = \begin{bmatrix} L_{11}h_1 \\ L_{21}h_1 + L_{22}h_2 \\ L_{32}h_2 + L_{33}h_3 \\ L_{43}h_3 + L_{44}h_4 \\ L_{54}h_4 + L_{55}h_5 \\ L_{65}h_5 + L_{66}h_6 \\ L_{76}h_6 + L_{77}h_7 \\ L_{87}h_7 + L_{88}h_8 \\ L_{98}h_8 + L_{99}h_9 \end{bmatrix} = \begin{bmatrix} r_1 \\ r_2 \\ r_3 \\ r_4 \\ r_5 \\ r_6 \\ r_7 \\ r_8 \\ r_9 \end{bmatrix}, \qquad \text{where,} \quad \mathbf{h} = \begin{bmatrix} h_1 \\ h_2 \\ h_3 \\ h_4 \\ h_5 \\ h_6 \\ h_7 \\ h_8 \\ h_9 \end{bmatrix}.$$

Therefore,

$$h_1 = \frac{r_1}{L_{11}}, \quad h_j = \frac{r_j - L_{j,j-1}h_{j-1}}{L_{jj}} \quad (2 \leq j \leq L). \tag{5.219}$$

And, from $U\mathbf{u} = \mathbf{h}$,

$$\begin{bmatrix} u_1 + U_{12}u_2 \\ u_2 + U_{23}u_3 \\ u_3 + U_{34}u_4 \\ u_4 + U_{45}u_5 \\ u_5 + U_{56}u_6 + U_{57}u_7 \\ u_6 + U_{67}u_7 \\ u_7 + U_{78}u_8 \\ u_8 + U_{89}u_9 \\ u_9 \end{bmatrix} = \begin{bmatrix} h_1 \\ h_2 \\ h_3 \\ h_4 \\ h_5 \\ h_6 \\ h_7 \\ h_8 \\ h_9 \end{bmatrix}.$$

Therefore,

$$\begin{cases} u_L = h_L, \\ u_j = h_j - U_{j,j+1}u_{j+1} \quad (j = L-1,\, L-2,\, \cdots,\, j_a+1), \\ u_{j_a} = h_{j_a} - U_{j_a,j_a+1}u_{j_a+1} - U_{j_a,j_a+2}u_{j_a+2}, \\ u_j = h_j - U_{j,j+1}u_{j+1} \quad (j = j_a-1,\, j_a-2,\, \cdots,\, 1), \end{cases} \tag{5.220}$$

where,

$$\mathbf{h}^{\mathrm{t}} = [h_1, h_2, h_3, \cdots, h_9].$$

Therefore,

$$h_1 = \frac{r_1}{L_{11}}, \quad h_j = \frac{r_j - L_{j,j-1}h_{j-1}}{L_{jj}} \quad (2 \leq j \leq L). \tag{5.221}$$

And, from $U\mathbf{u} = \mathbf{h}$,

$$\begin{bmatrix} u_1 + U_{12}u_2 \\ u_2 + U_{23}u_3 \\ u_3 + U_{34}u_4 \\ u_4 + U_{45}u_5 \\ u_5 + U_{56}u_6 + U_{57}u_7 \\ u_6 + U_{67}u_7 \\ u_7 + U_{78}u_8 \\ u_8 + U_{89}u_9 \\ u_9 \end{bmatrix} = \begin{bmatrix} h_1 \\ h_2 \\ h_3 \\ h_4 \\ h_5 \\ h_6 \\ h_7 \\ h_8 \\ h_9 \end{bmatrix}.$$

Therefore,

$$\begin{cases} u_L = h_L, \\ u_j = h_j - U_{j,j+1}u_{j+1} \quad (j = L-1, L-2, \cdots, j_a + 1), \\ u_{j_a} = h_{j_a} - U_{j_a, j_a+1}u_{j_a+1} - U_{j_a, j_a+2}u_{j_a+2}, \\ u_j = h_j - U_{j,j+1}u_{j+1} \quad (j = j_a - 1, j_a - 2, \cdots, 1). \end{cases} \tag{5.222}$$

Here, we introduce new variables p_j $(1 \le j \le L-1)$ and q to simplify the expression as follows:

$$\begin{cases} p_1 = U_{12} = \dfrac{c_1}{b_1}, \\ p_j = U_{j,j+1} = \dfrac{c_j}{b_j - a_j p_{j-1}} \quad (j = 2, 3, \cdots, j_a), \\ q = U_{j_a, j_a+2} = \dfrac{d_{j_a}}{b_{j_a} - a_{j_a} p_{j_a-1}}, \\ p_{j_a+1} = U_{j_a+1, j_a+2} = \dfrac{c_{j_a+1} - a_{j_a+1}q}{b_{j_a+1} - a_{j_a+1} p_{j_a}}, \\ p_j = U_{j,j+1} = \dfrac{c_j}{b_j - a_j p_{j-1}} \quad (j = j_a + 2, j_a + 3, \cdots, L-1), \end{cases}$$

then,

$$\begin{cases} h_1 = \dfrac{r_1}{b_1}, \\ h_j = \dfrac{r_j - h_{j-1}a_j}{b_j - a_j p_{j-1}} \quad (j = 2, 3, \cdots, L). \end{cases} \tag{5.223}$$

Finally, we get

$$\begin{cases} u_L = h_L, \\ u_j = h_j - p_j u_{j+1}, \quad (j = L-1, L-2, \cdots, j_a + 1), \\ u_{j_a} = h_{j_a} - p_{j_a} u_{j_a+1} - q u_{j_a+2}, \\ u_j = h_j - p_j u_{j+1}, \quad (j = j_a - 1, j_a - 2, \cdots, 1). \end{cases} \tag{5.224}$$

Where a_j $(2 \leq j \leq L)$, b_j $(1 \leq j \leq L)$, c_j $(1 \leq j \leq L-1)$, r_j $(1 \leq j \leq L)$, and d_{j_a} are given by the following equations:

- In the case of $(1 \leq j \leq j_a - 1)$,

$$a_j = c_j = \frac{\rho_j}{(\Delta\rho_a)^2},\ b_j = -2\left[\frac{\rho_j}{(\Delta\rho_a)^2} + \frac{\mathcal{B}(\rho_j, u_j^n)}{\Delta\beta}\right],$$
$$r_j = -2\frac{\mathcal{B}(\rho_j, u_j^n)}{\Delta\beta}u_j^n. \tag{5.225}$$

However, a_1 is a coefficient that is not actually used.

- In the case of $j = j_a$,

$$a_j = \frac{\rho_j}{(\Delta\rho_a)^2},\ b_j = -\left[\frac{\rho_j}{(\Delta\rho_a)^2}(2 - \hat{c}_{u1}) + 2\frac{\mathcal{B}(\rho_j, u_j^n)}{\Delta\beta}\right],$$
$$c_j = \frac{\rho_j}{(\Delta\rho_a)^2}\hat{c}_{u2},\ d_j = \frac{\rho_j}{(\Delta\rho_a)^2}\hat{c}_{u3},\ r_j = -2\frac{\mathcal{B}(\rho_j, u_j^n)}{\Delta\beta}u_j^n. \tag{5.226}$$

- In the case of $(j_a + 1 \leq j \leq L - 1)$,

$$a_j = c_j = \frac{\rho_j}{(\Delta\rho_b)^2},\ b_j = -2\left[\frac{\rho_j}{(\Delta\rho_b)^2} + \frac{\mathcal{B}(\rho_j, u_j^n)}{\Delta\beta}\right],$$
$$r_j = -2\frac{\mathcal{B}(\rho_j, u_j^n)}{\Delta\beta}u_j^n. \tag{5.227}$$

- In the case of $j = L$,

$$a_j = \frac{\rho_j}{(\Delta\rho_b)^2},\ b_j = -2\left[\frac{\rho_j}{(\Delta\rho_b)^2} + \frac{\mathcal{B}(\rho_j, u_j^n)}{\Delta\beta}\right],$$
$$r_j = -2\frac{\mathcal{B}(\rho_j, u_j^n)}{\Delta\beta}u_j^n - \frac{\rho_j}{(\Delta\rho_b)^2}u_S(\rho_{L+1}). \tag{5.228}$$

For Corrector, these coefficients are given by

- In the case of $(1 \leq j \leq j_a - 1)$,

$$a_j = c_j = \frac{\rho_j}{(\Delta\rho_a)^2},\ b_j = -2\left[\frac{\rho_j}{(\Delta\rho_a)^2} + \frac{\mathcal{B}(\rho_j, u_j^{n+1/2})}{\Delta\beta}\right],$$
$$r_j = -\frac{\rho_j}{(\Delta\rho_a)^2}u_{j+1}^n + 2\left[\frac{\rho_j}{(\Delta\rho_a)^2} - \frac{\mathcal{B}(\rho_j, u_j^{n+1/2})}{\Delta\beta}\right]u_j^n$$
$$-\frac{\rho_j}{(\Delta\rho_a)^2}u_{j-1}^n. \tag{5.229}$$

- In the case of $j = j_a$,

$$a_j = \frac{\rho_j}{(\Delta\rho_a)^2},\ b_j = -\left[\frac{\rho_j}{(\Delta\rho_a)^2}(2 - \hat{c}_{u1}) + 2\frac{\mathcal{B}(\rho_j, u_j^{n+1/2})}{\Delta\beta}\right],$$

$$c_j = \frac{\rho_j}{(\Delta\rho_a)^2}\hat{c}_{u2},\ d_j = \frac{\rho_j}{(\Delta\rho_a)^2}\hat{c}_{u3}$$

$$r_j = -\frac{\rho_j}{(\Delta\rho_a)^2}\widetilde{u}_{j+1}^n + 2\left[\frac{\rho_j}{(\Delta\rho_a)^2} - \frac{\mathcal{B}(\rho_j, u_j^{n+1/2})}{\Delta\beta}\right]u_j^n$$
$$-\frac{\rho_j}{(\Delta\rho_a)^2}u_{j-1}^n, \tag{5.230}$$

where,

$$\widetilde{u}_{j+1}^n = \hat{c}_{u1}u_j^n + \hat{c}_{u2}u_{j+1}^n + \hat{c}_{u3}u_{j+2}^n.$$

- In the case of $(j_a + 1 \le j \le L - 1)$,

$$a_j = c_j = \frac{\rho_j}{(\Delta\rho_b)^2},\ b_j = -2\left[\frac{\rho_j}{(\Delta\rho_b)^2} + \frac{\mathcal{B}(\rho_j, u_j^{n+1/2})}{\Delta\beta}\right],$$

$$r_j = -\frac{\rho_j}{(\Delta\rho_b)^2}u_{j+1}^n + 2\left[\frac{\rho_j}{(\Delta\rho_b)^2} - \frac{\mathcal{B}(\rho_j, u_j^{n+1/2})}{\Delta\beta}\right]u_j^n$$
$$-\frac{\rho_j}{(\Delta\rho_b)^2}u_{j-1}^n. \tag{5.231}$$

- In the case of $j = L$,

$$a_j = \frac{\rho_j}{(\Delta\rho_b)^2},\ b_j = -2\left[\frac{\rho_j}{(\Delta\rho_b)^2} + \frac{\mathcal{B}(\rho_j, u_j^{n+1/2})}{\Delta\beta}\right],$$

$$r_j = -2\frac{\rho_j}{(\Delta\rho_b)^2}u_S(\rho_{L+1}) + 2\left[\frac{\rho_j}{(\Delta\rho_b)^2} - \frac{\mathcal{B}(\rho_j, u_j^{n+1/2})}{\Delta\beta}\right]u_j^n$$
$$-\frac{\rho_j}{(\Delta\rho_b)^2}u_{j-1}^n. \tag{5.232}$$

5.10.2.6 Solutions for $R \le j \le J$

Here, the system of linear equations to be solved is exactly the same as the system of linear equations described in Section 5.8.3 if $\Delta\rho = \Delta\rho_b$, the solutions are given by Eqs. (5.173) and (5.174).

Chapter 6

Calculation Procedure of SCOZA

In the previous chapter, we have explained the equations required to numerically calculate the excess internal energy $u(\rho, \beta)$ of a fluid. In this chapter, we explain the calculation procedure for actually finding u using these equations.

6.1 Method to Determine the Potential Tail

It was shown in Chapter 4 that there are many intermolecular interactions that reproduce the same equation of state. Here, taking $\phi_{c1}(r)$ as an example among them, the procedure for calculating the thermodynamic properties of the fluid is specifically described. First, how to determine the function form of $\phi_{c1}(r)$ is described. In this book, we consider that water molecules are made of hard spheres with potential tails consisting of soft repulsive force and attractive force, and explore what shape tails reproduce well the thermodynamic properties of water. It is impossible to know the shape of such tails in advance. Therefore, it is required that work is done on several guessed tails to find the optimal interaction that reproduces the experimental results well while adjusting the shape. This is similar to exploring an ideal airplane shape that minimizes air resistance.

The Physics of Liquid Water
Makoto Yasutomi

ISBN 978-981-4877-25-1 (Hardcover), 978-1-003-05616-4 (eBook)
www.jennystanford.com

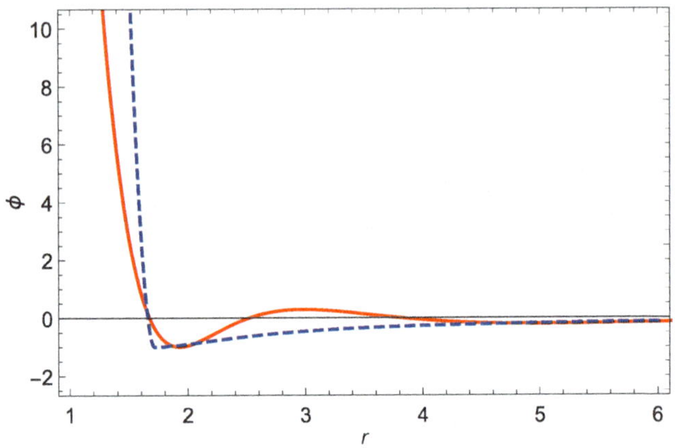

Figure 6.1 A method to express the interaction between water molecules with multi-Yukawa potential. The guessed shape of the interaction between water molecules is shown by a broken line. Under the condition that the potential tail is made up of five Yukawa terms, the shape that best reproduces the dashed line is $\phi_{c1}(r)$, shown in red.

As an example, the guessed tail shape is shown as a broken line in Fig. 6.1. Under the condition that tail ϕ_{c1} is made of five Yukawa terms, the function $\phi_{c1}(r)$ that best fits the broken line is shown in red.

To confirm whether the potential tail $\phi_{c1}(r)$ obtained in this way can reproduce the thermodynamic properties of water with high accuracy, it is necessary to calculate all the required physical quantities using SCOZA and thermodynamic related equations. Therefore, it is necessary to repeat the trial calculation several times while adjusting the shape of the potential tail that generates the soft repulsive force and attractive force to find the optimum shape. The potential $\phi_{c1}(r)$ is a tail obtained as a result of repeating such a trial calculation several times.

6.2 $\mathcal{K}^{(1)}(\rho)$ and $z_1(\rho)$

The direct correlation function given by Eq. (5.79) of the hard sphere fluid is obtained by [Pini (1998)]. The quantities $\mathcal{K}^{(1)}(\rho)$ and $z_1(\rho)$

included in the formula are given by the following equation:

$$\mathcal{K}^{(1)} = \frac{2(z_1+2)^2\sigma_1^2}{3\eta z_1^2} V_0 \left(\frac{V_1}{V_0} - \alpha_1\right)^2,$$

$$z_1 = \frac{2}{q-f^2}\left[(V_0+f^2-q)f + \sqrt{(V_0+f^2-q)V_0 q}\right],$$

where,

$$q = \frac{(1+2\eta)^2}{(1-\eta)^2}, \quad f = \frac{1-\eta}{\sqrt{\chi_{\text{red}}}}, \quad V_0 = 6\eta y_0 - f^2 + 1,$$

$$y_0 = \frac{1}{4\eta}\left(\frac{\beta p}{\rho} - 1\right).$$

Here χ_{red} is the reduced isothermal compressibility of the hard sphere fluid and is given by the following equation (see Fig. 6.2):

$$\frac{1}{\chi_{\text{red}}} = \frac{1+4\eta+4\eta^2-4\eta^3+\eta^4}{(1-\eta)^4}.$$

The quantity $\beta p/\rho$ is given by Carnahan–Starling's empirical formula (5.76) (see blue line in Fig. 5.3). Also,

$$\frac{V_1}{V_0} = 2 - \sqrt{q} - \frac{1}{2V_0\sqrt{q}}\left[(V_0+f^2-q)(V_0+f^2) + \frac{z_1^2}{4}(q-f^2)\right],$$

$$\sigma_1 = \frac{1}{2z_1}\left(\frac{z_1-2}{z_1+2} + e^{-z_1}\right), \quad \tau_1 = \frac{1}{2z_1}\left(\frac{z_1^2+2z_1-4}{4+2z_1-z_1^2} + e^{-z_1}\right),$$

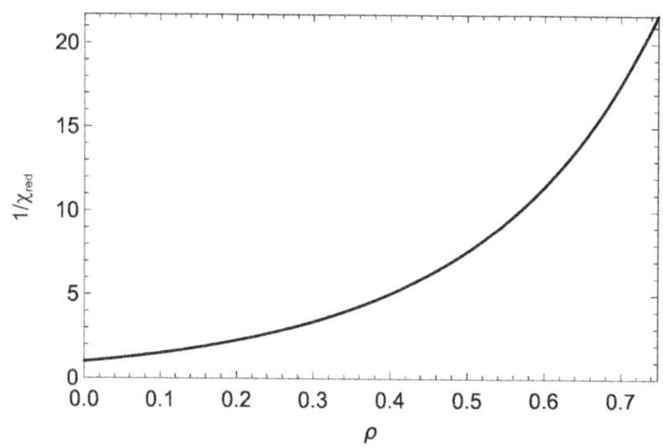

Figure 6.2 $1/\chi_{\text{red}}(\rho)$.

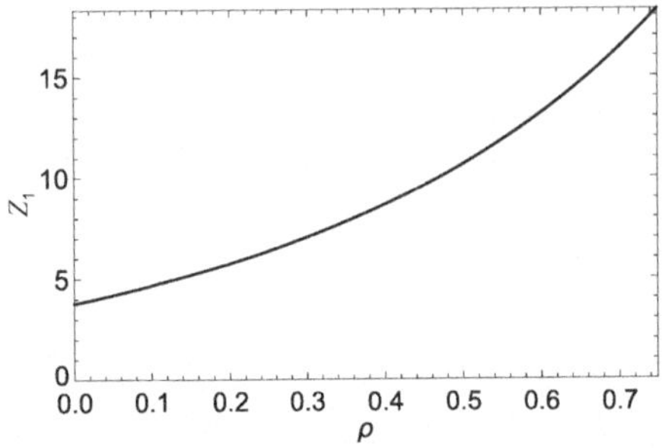

Figure 6.3 $z_1(\rho)$.

$$\alpha_1 = \frac{(4 + 2z_1 - z_1^2)\tau_1}{2(2 + z_1)\sigma_1}.$$

The coefficients $z_1(\rho)$ and $\mathcal{K}^{(1)}(\rho)$ calculated from the above equations are numerically unstable in the region where ρ is sufficiently small, and give occasionally incorrect values. Therefore, in such a region, the quantities $z_1(\rho)$ and $\mathcal{K}^{(1)}(\rho)$ should be expanded to a power series of $\eta = \pi\rho/6$ as shown below.

$$z_1 = 3.79129 + 15.2947\eta + 27.1069\eta^2 + 36.0378\eta^3$$

$$+51.013\eta^4 + 54.0043\eta^5 + 78.0014\eta^6.$$

$$\mathcal{K}^{(1)} = 0.5\eta^2 + 2.88618\eta^3 + 5.84461\eta^4 + 5.27577\eta^5 + 4.00947\eta^6.$$

In the region of $\eta < 10^{-2}$, we use the above power series expansion by η for z_1 and $\mathcal{K}^{(1)}$. The obtained $z_1(\rho)$ and $\mathcal{K}^{(1)}(\rho)$ are shown in Figs. 6.3 and 6.4, respectively. The density increment was set to $\Delta\rho = 10^{-4}$.

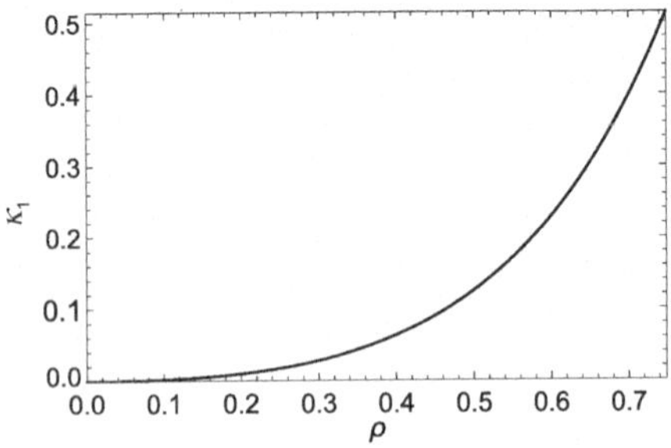

Figure 6.4 $\mathcal{K}_1(\rho)$.

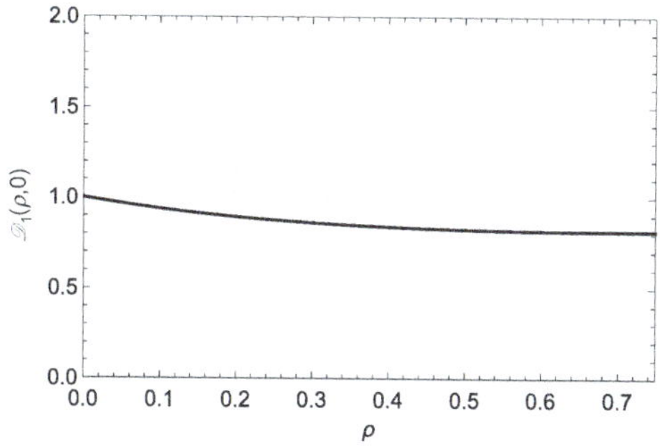

Figure 6.5 $\mathscr{D}_1(\rho, 0)$.

6.2.1 Calculation of $u(\rho, \beta = 0)$

When $\beta = 0$ $(T \to \infty)$, the intermolecular interaction $\phi_{c1}(r)$ is negligible compared to the thermal energy per molecule. In this case, the fluid can be regarded as a hard sphere fluid. As explained in Section 5.7.2, for a hard sphere fluid, $\mathscr{D}_n = 0$ $(2 \leq n \leq N = 6)$. Therefore, the values of $(\mathscr{G}_1, \mathscr{D}_1)$ can be obtained from Eqs. (5.128) and (5.129) using the Newton–Raphson method. The values of $\mathscr{G}_n$ $(2 \leq n \leq 6)$ can be obtained from Eq. (5.130). The value of $\mathscr{D}_1(\rho, 0)$ obtained in this way is shown in Fig. 6.5.

The solutions $\mathscr{G}_n$ $(1 \leq n \leq 6)$ are shown in Fig. 6.6. Substituting $\mathscr{G}_n(\rho, 0)$ $(2 \leq n \leq 6)$ into Eq. (5.123) gives $u_{\text{HS}} = u(\rho, 0)$. The result is shown in Fig. 6.7. Substituting u_{HS} into the diffusion equation

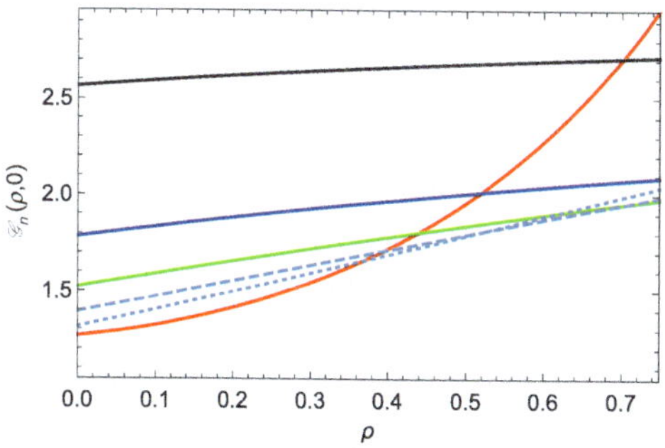

Figure 6.6 $\mathscr{G}_n(\rho, 0)$ $(1 \leq n \leq 6)$. Red line: $\mathscr{G}_1(\rho, 0)$, Black line: $\mathscr{G}_2(\rho, 0)$, Blue line: $\mathscr{G}_3(\rho, 0)$, Green line: $\mathscr{G}_4(\rho, 0)$, Dashed line: $\mathscr{G}_5(\rho, 0)$, Dotted line: $\mathscr{G}_6(\rho, 0)$.

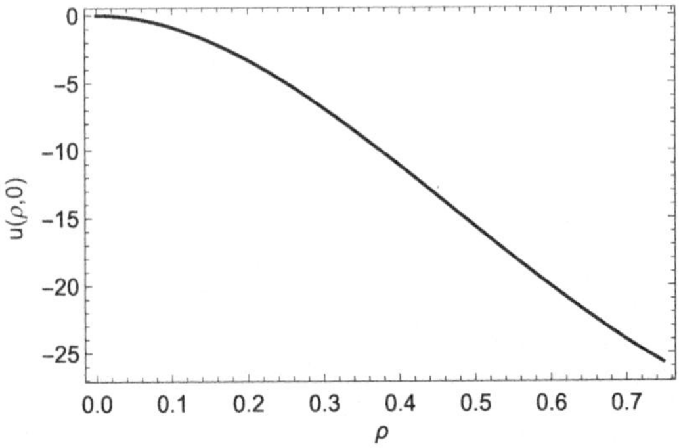

Figure 6.7 $u_{\mathrm{HS}} = u(\rho, 0)$.

(5.124), the quantity $u(\rho, \beta_1)$ at the next temperature step $\beta_1 = \Delta\beta_1$ can be obtained.

Diffusion coefficient $\mathcal{B}(\rho, u_{\mathrm{HS}})$ is obtained from Eq. (5.125). The resulting $\mathcal{B}(\rho, u_{\mathrm{HS}})$ is shown in Fig. 6.8. The methods to find A, $\partial A/\partial x_n$, and $\partial x_n/\partial u$ included in that equation are explained in Appendix D and Appendix E. In the density region of $\rho > \rho_{\mathrm{J}}$, there is no stable solution of the diffusion equation (5.124) because $\mathcal{B}(\rho, u_{\mathrm{HS}})$ is negative. So we set the boundary density to $\rho = \rho_{\mathrm{J}}$ (see Table 5.1). Therefore, the density region that should be considered is given by $0 < \rho \leq \rho_{\mathrm{J}}$. In this case, the boundary condition (5.136) is given by the following equation:

$$\Delta^2 u^0 = 7.16947 \times 10^{-7} \quad \text{at } \rho = \rho_{\mathrm{J}} = 0.7487.$$

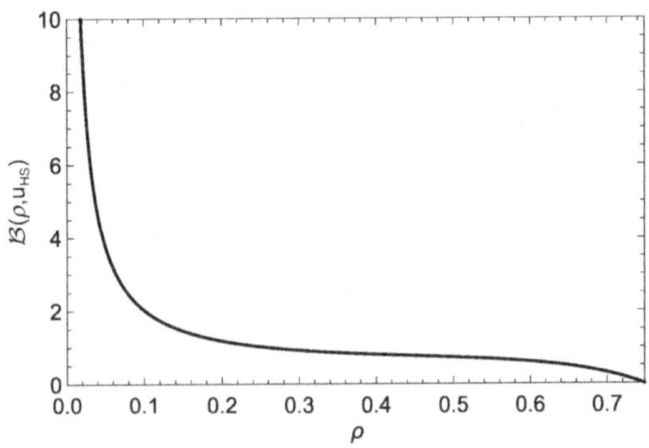

Figure 6.8 $\mathcal{B}(\rho, u_{\mathrm{HS}})$. $\mathcal{B}(\rho_1, u_{\mathrm{HS}}) = 176176.0$ and $\mathcal{B}(\rho_{\mathrm{J}}, u_{\mathrm{HS}}) = 5.77 \times 10^{-5}$.

6.2.2 Calculation up to Critical Temperature T_c

You can use $\mathscr{D}_n(\rho, \beta = 0)$, $\mathscr{G}_n(\rho, \beta = 0)$ $(1 \le n \le N)$, $u_{HS}(\rho) = u(\rho, \beta = 0)$, and $\mathcal{B}(\rho, u_{HS})$ as initial values. If you solve the diffusion equation (5.124) using the method described in Section 5.8, the excess internal energy $u(\rho, \beta_k)$ for any temperature step β_k can be obtained. The behavior of $\mathcal{B}(\rho, u)$ obtained in this way near the critical point (ρ_c, β_c) is shown in Fig. 6.9. You can see how $\mathcal{B}(\rho, u)$ suddenly approaches 0 around $\rho = \rho_c = 0.0841$ as β approaches β_c.

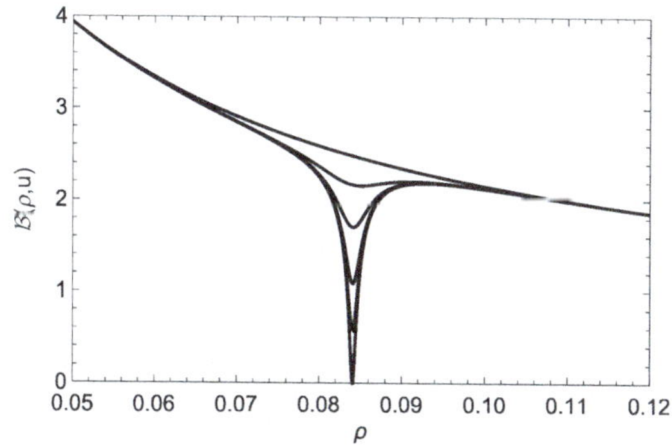

Figure 6.9 The behavior of $\mathcal{B}(\rho, u)$ near the critical point (ρ_c, β_c). The curves represent the isotherms of $\mathcal{B}(\rho, u)$ for $\beta = 0.1025, 0.10776, 0.10787, 0.107886, 0.107891$ and $\beta_c = 0.107891313$ from the top to the bottom, respectively.

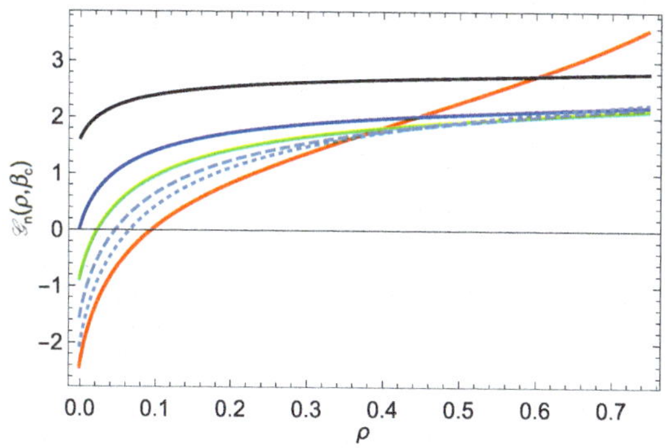

Figure 6.10 $\mathscr{G}_n(\rho, \beta_c)$. Red line: $\mathscr{G}_1(\rho, \beta_c)$, Black line: $\mathscr{G}_2(\rho, \beta_c)$, Blue line: $\mathscr{G}_3(\rho, \beta_c)$, Green line: $\mathscr{G}_4(\rho, \beta_c)$, Dashed line: $\mathscr{G}_5(\rho, \beta_c)$, Dotted line: $\mathscr{G}_6(\rho, \beta_c)$.

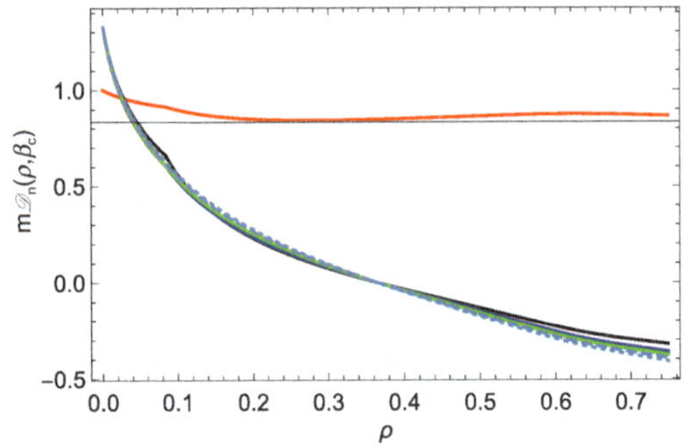

Figure 6.11 $\mathscr{D}_n(\rho, \beta_c)$. Red line: $\mathscr{D}_1(\rho, \beta_c)$, Black line: $\mathscr{D}_2(\rho, \beta_c)$, Blue line: $\mathscr{D}_3(\rho, \beta_c)$, Green line: $\mathscr{D}_4(\rho, \beta_c)$, Dashed line: $\mathscr{D}_5(\rho, \beta_c)$, Dotted line: $\mathscr{D}_6(\rho, \beta_c)$.

The solutions $\mathscr{G}_n(\rho, \beta_c)$ and $\mathscr{D}_n(\rho, \beta_c)$ $(1 \leq n \leq 6)$ at the critical temperature $\beta = \beta_c$ are shown in Figs. 6.10 and 6.11.

6.2.3 Calculation below Critical Temperature

As explained in Section 5.7.3, below the critical temperature T_c, an unstable gas–liquid coexistence region appears in the middle density region where $\mathcal{B}(\rho, u) < 0$. Gas is stable on the lower density side and liquid is stable on the higher density side than that. In order to obtain a solution in the stable region, a spinodal curve satisfying the boundary condition $A(\rho_{S1}, \beta) = A(\rho_{S2}, \beta) = 0$ [see Eq. (5.131)] must be solved. The values of $u(\rho_{Si})$ $(i = 1, 2.)$ on the spinodal

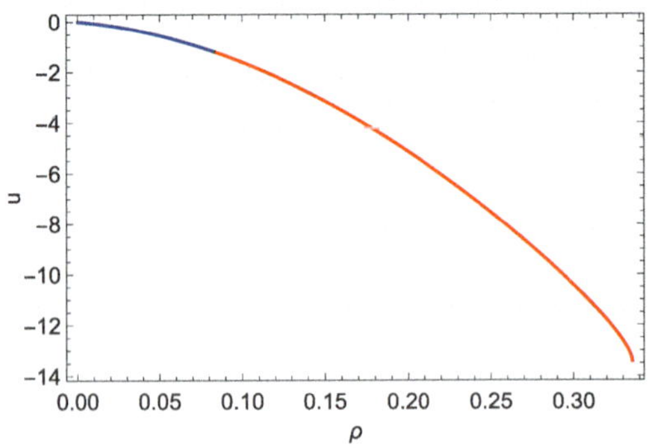

Figure 6.12 $u(\rho_{Si})$ $(i = 1, 2)$. Blue line: $u(\rho_{S1})$, Red line: $u(\rho_{S2})$.

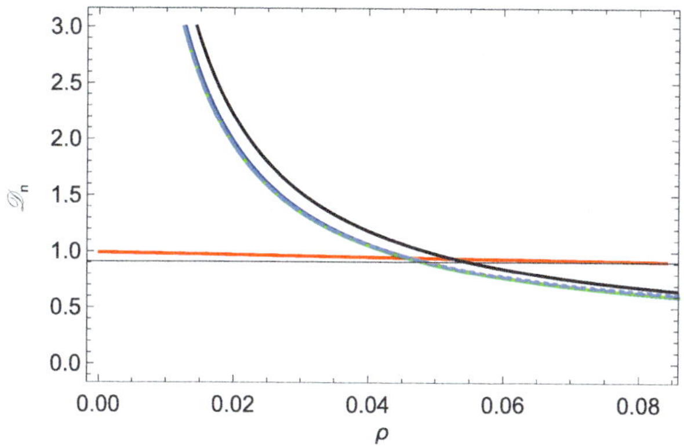

Figure 6.13 $\mathscr{D}_n(\rho_{S1})$ $(1 \leq n \leq 6)$. Red line: $\mathscr{D}_1(\rho_{S1})$, Black line: $\mathscr{D}_2(\rho_{S1})$, Blue line: $\mathscr{D}_3(\rho_{S1})$, Green line: $\mathscr{D}_4(\rho_{S1})$, Dashed line: $\mathscr{D}_5(\rho_{S1})$, Dotted line: $\mathscr{D}_6(\rho_{S1})$,

curve are shown in Fig. 6.12. The blue line on the lower density side than $\rho_c = 0.0841$ represents $u(\rho_{S1})$, and the red line on the higher density side represents $u(\rho_{S2})$. The density ρ_{S2} has an upper limit value of 0.3356, and the value of $u(\rho_{S2})$ diverges in the density region $\rho_{S2} > 0.3356$.

The values of $\mathscr{D}_n(\rho_{S1})$ on the spinodal curve are shown in Fig. 6.13, $\mathscr{D}_n(\rho_{S2})$ in Fig. 6.14, $\mathscr{G}_n(\rho_{S1})$ in Fig. 6.15, and $\mathscr{G}_n(\rho_{S2})$ in Fig. 6.16, where $(1 \leq n \leq 6)$.

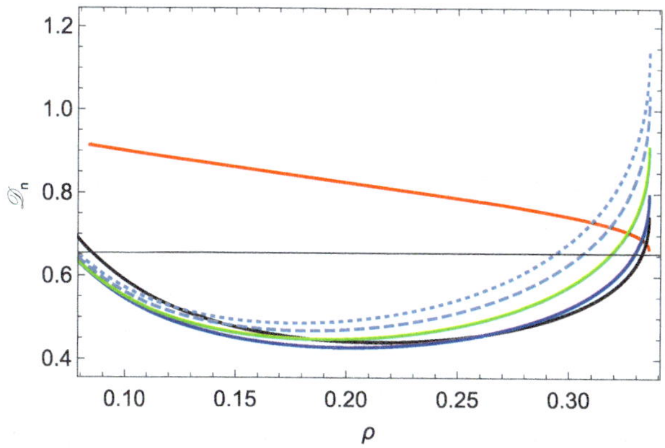

Figure 6.14 The same as Fig. 6.13, but for $\mathscr{D}_n(\rho_{S2})$ $(1 \leq n \leq 6)$.

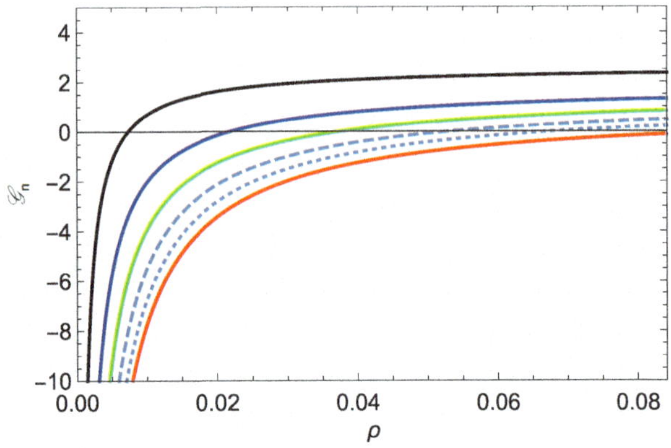

Figure 6.15 The same as Fig. 6.13, but for $\mathscr{G}_n(\rho_{S1})$ $(1 \leq n \leq 6)$.

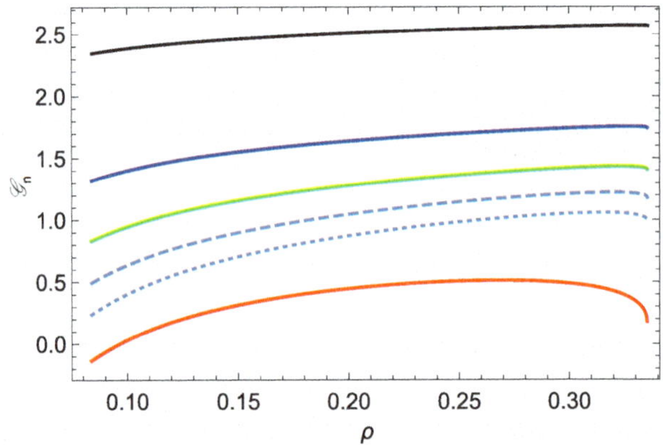

Figure 6.16 The same as Fig. 6.13, but for $\mathscr{G}_n(\rho_{S2})$ $(1 \leq n \leq 6)$.

6.2.4 Calculations of Gas Phase Water below the Critical Temperature

Calculations for steam below the critical temperature can be made using the method described in Section 5.8.2. First, $[\overrightarrow{\mathscr{G}}(\rho, \beta_c), \overrightarrow{\mathscr{D}}(\rho, \beta_c)]$ and $[\overrightarrow{\mathscr{G}}(\rho_{S1}), \overrightarrow{\mathscr{D}}(\rho_{S1})]$ are used as approximate solutions for the Newton–Raphson method to find solutions for the next temperature step $\beta = \beta_c + \Delta\beta$. Similarly, if the solutions for this temperature are used as approximate solutions for the next temperature step, solutions for each temperature step can be

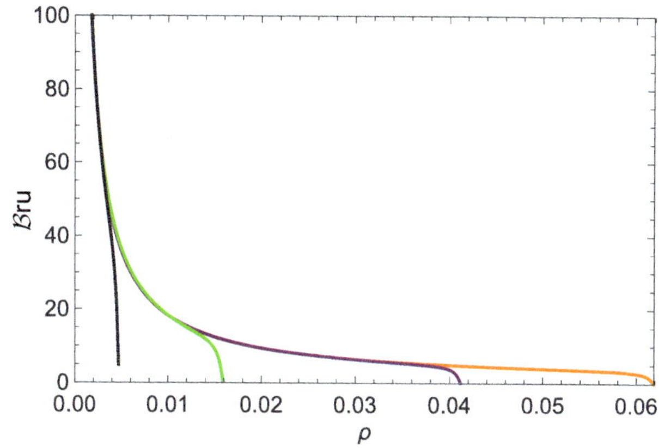

Figure 6.17 $\mathcal{B}(\rho, u)$ of gas phase water at typical temperatures $\beta = 0.1154$ (orange line), 0.1414 (purple line), 0.2684 (green line), and 0.5954 (black line).

obtained one after another. The solutions $\mathcal{B}(\rho, u)$ and $u(\rho, \beta)$ obtained in this way are shown for representative temperatures $\beta = 0.1154$, 0.1414, 0.2684, and 0.5954 in Figs. 6.17 and 6.18, respectively.

For the same representative temperatures as above, the solutions $\mathscr{D}_n(\rho, \beta)$ and $\mathscr{G}_n(\rho, \beta)$ $(1 \leq n \leq 6)$ are also exhibited in Figs. G.1, G.2, G.3, G.4, G.5, G.6, G.7, and G.8, respectively, in Section G.1 in Appendix G.

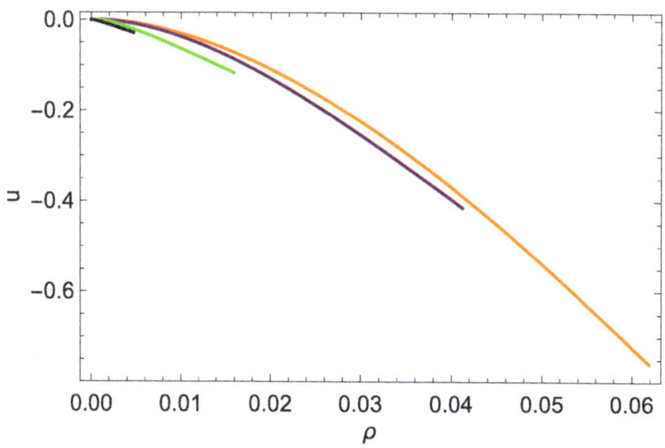

Figure 6.18 The same as Fig. 6.17, but for $u(\rho, \beta)$.

6.2.5 Calculations of Liquid Phase Water below the Critical Temperature

Liquid water below the critical temperature can be calculated by the method described in Section 5.8.3. The solutions of Eq. (5.124) are obtained in the same way as the calculation method for vapor phase water, using the solutions of the previous temperature step as approximate solutions.

The solutions $\mathcal{B}(\rho, u)$ and $u(\rho, \beta)$ are shown for the same representative temperatures as previous subsection, respectively, in Figs. 6.19 and 6.20.

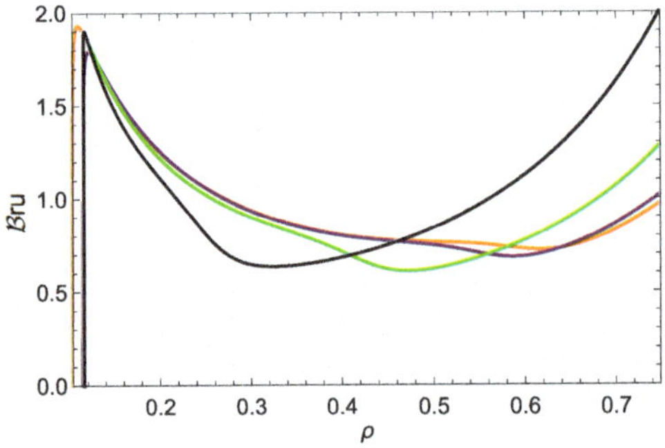

Figure 6.19 The same as Fig. 6.17, but for liquid water.

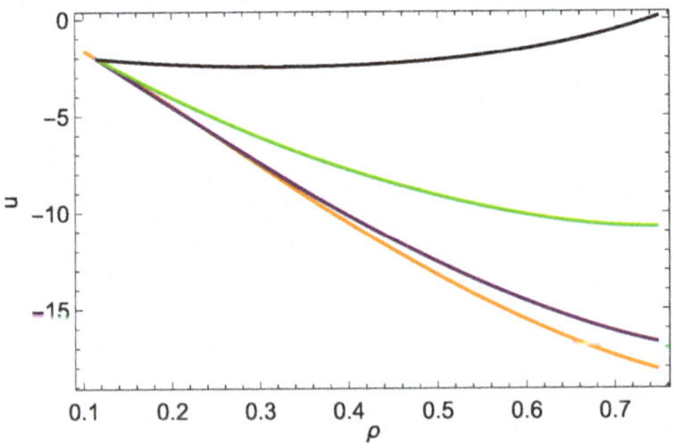

Figure 6.20 The same as Fig. 6.18, but for liquid water.

The solutions $\mathcal{D}_n(\rho, \beta)$ and $\mathcal{G}_n(\rho, \beta)$ $(1 \leq n \leq 6)$ are exhibited in Figs. G.9, G.10, G.11, G.12, G.13, G.14, G.15, and G.16, respectively, in Section G.2 in Appendix G.

Chapter 7

Pressure and Chemical Potential

Pressure and chemical potential can be derived from both energy equation and compressibility equation. If the theory is rigorous, the pressure and chemical potential obtained from these two different equations should each be in perfect agreement. However, since some approximation has been introduced into all existing fluid theories, they do not match perfectly. SCOZA is a theory that is constructed so that the two match each other. Actually, when numerical calculations are performed, it can be seen that the two match with high accuracy. Strictly speaking, however, there is a slight shift between the two, which creates, for example, a subtle contradiction when determining the liquefaction point and the vaporization point. In order to resolve this contradiction, it is necessary to adopt both calculation methods well. As a result of carrying out such a careful calculation method, it was found that there is a region where the liquefaction point and the vaporization point are absent just below the critical point. This chapter details such matters.

7.1 Pressure Derived from Energy Equation

Since $u(\rho, \beta)$ was obtained in the previous chapter, if we integrate both sides of the energy equation Eq. (1.17) with respect to β, the

The Physics of Liquid Water
Makoto Yasutomi

ISBN 978-981-4877-25-1 (Hardcover), 978-1-003-05616-4 (eBook)
www.jennystanford.com

state equation $\beta p/\rho$ can be obtained. If $\beta p(\rho, \beta_*)$ is obtained for a certain temperature $\beta = \beta_*$, integrating both sides of (1.17) for β in both ranges from β_* to β_i and β_{i+1} gives

$$\beta p(\rho, \beta_i) = \beta p(\rho, \beta_*) + \int_{\beta_*}^{\beta_i} \left[-u + \rho \left(\frac{\partial u}{\partial \rho} \right)_\beta \right] d\beta.$$

$$\beta p(\rho, \beta_{i+1}) = \beta p(\rho, \beta_*) + \int_{\beta_*}^{\beta_{i+1}} \left[-u + \rho \left(\frac{\partial u}{\partial \rho} \right)_\beta \right] d\beta.$$

From the above two equations, the following equation is obtained using the trapezoidal formula for integration:

$$\beta p(\rho, \beta_{i+1}) = \beta p(\rho, \beta_i) + \frac{\Delta\beta}{2} \left\{ \left[-u + \rho \left(\frac{\partial u}{\partial \rho} \right)_\beta \right]_{\beta_i} + \left[-u + \rho \left(\frac{\partial u}{\partial \rho} \right)_\beta \right]_{\beta_{i+1}} \right\},$$

where, $\Delta\beta = \beta_{i+1} - \beta_i$.

The differential coefficient $(\partial u/\partial \rho)_\beta$ is obtained as follows. Here, a case will be described in which the step sizes of the density as described in Section 5.9 are different between those for $j < j_a$ and $j > j_a$. For $j \neq j_a$, it is given by

$$\left(\frac{\partial u}{\partial \rho} \right)_\beta (\rho_0 = 0, \beta) = \frac{u(\rho_1, \beta) - u(0, \beta)}{\rho_1} \qquad (j = 0).$$

$$\left(\frac{\partial u}{\partial \rho} \right)_\beta (\rho_j, \beta) = \frac{u(\rho_{j+1}, \beta) - u(\rho_{j-1}, \beta)}{\rho_{j+1} - \rho_{j-1}}$$

for $(1 \leq j < j_a$ or $j_a < j \leq J - 1)$,

$$\left(\frac{\partial u}{\partial \rho} \right)_\beta (\rho_J, \beta) = \frac{u(\rho_J, \beta) - u(\rho_{J-1}, \beta)}{\rho_J - \rho_{J-1}} \qquad (j = J).$$

When $j = j_a$, it is given by the following formula (see Section 5.10.1.2):

$$\left(\frac{\partial u}{\partial \rho} \right)_\beta (\rho_{j_a}, \beta) = \frac{\tilde{u}(\rho_{j_a+1}, \beta) - u(\rho_{j_a-1}, \beta)}{2(\rho_{j_a} - \rho_{j_a-1})}.$$

where, $\tilde{u}(\rho_{j_a+1}, \beta) = u(\rho_{j_a} + \Delta\rho_a, \beta) = \hat{c}_{u1}\hat{u}_1 + \hat{c}_{u2}\hat{u}_2 + \hat{c}_{u3}\hat{u}_3$.

7.2 Pressure Derived from Compressibility Equation

In Section 7.1, the pressure is obtained from the energy equation, but here the pressure is derived from the compressibility equation. From Eqs. (1.21) and (5.122), we get

$$\left(\frac{\partial \beta p}{\partial \rho}\right)_\beta = \left[\frac{A(\rho, \beta)}{2\pi}\right]^2. \tag{7.1}$$

Integrating the above equation for ρ in both ranges from ρ_* to ρ_j and ρ_{j+1} yields

$$\beta p(\rho_j, \beta) = \beta p(\rho_*, \beta) + \int_{\rho_*}^{\rho_j} \left[\frac{A(\rho, \beta)}{2\pi}\right]^2 d\rho.$$

$$\beta p(\rho_{j+1}, \beta) = \beta p(\rho_*, \beta) + \int_{\rho_*}^{\rho_{j+1}} \left[\frac{A(\rho, \beta)}{2\pi}\right]^2 d\rho.$$

From the above two equations and the trapezoidal formula for integration, the following equation is obtained:

$$\beta p(\rho_{j+1}, \beta) = \beta p(\rho_j, \beta) + \frac{\rho_{j+1} - \rho_j}{8\pi^2}\left[A(\rho_j, \beta)^2 + A(\rho_{j+1}, \beta)^2\right].$$

For any β, if $\beta p(\rho_*, \beta)$ is given for a certain density ρ_*, then using the above formula, $\beta p(\rho_j, \beta)$ is obtained for any density ρ_j. If you use the energy equation, you can get p on the spinodal line for any β_i, so if you set it as the initial value, you can use the above pressure equation derived from the compressibility equation to get $\beta p(\rho_j, \beta_i)$ at any point (ρ_j, β_i).

If you use $p(\rho_{S_1}, \beta_i)$ as the initial value in the gas phase region below the critical temperature $(\beta > \beta_c)$, the quantity $\beta p(\rho_j, \beta_i)$ is obtained for any step in the range $\rho_1 < \rho_j \leq \rho_{S_1}$. Here, ρ_{S_1} represents the density of the spinodal point on the low-density side. In the liquid phase region, if $p(\rho_{S_2}, \beta_i)$ is used as an initial value, the value of $\beta p(\rho_j, \beta_i)$ is obtained for any step in the region of $\rho_{S_2} < \rho_j \leq \rho_J$. Where ρ_{S_2} is the density of the spinodal point on the high-density side, and ρ_J is the density at the boundary point on the high-density side.

7.3 In the Case of $T \geq T_c$

Put $\beta_* = 0$ on the initial value $\beta p(\rho, \beta_*)$, apply Carnahan–Staring formula (5.76), and set $\beta p(\rho, \beta_* = 0) = (1+\eta+\eta^2-\eta^3)/\Delta^3$. Using the pressure equation derived from the energy equation obtained in the previous section, we can find $\beta p(\rho_j, \beta_i)$ for any step between β_1 and critical temperature β_c for any density ρ_j.

7.3.1 Spinodal Curve

If only the pressure equation obtained from the energy equation is used, inconvenience arises in obtaining the liquefaction point and the vaporization point. In order to solve this problem, it is

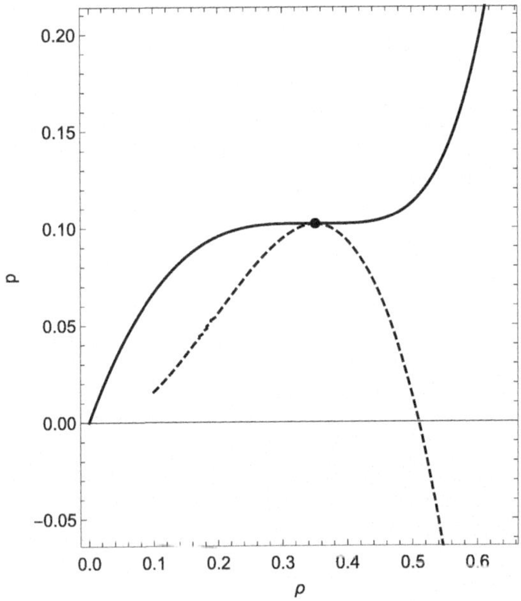

Figure 7.1 Spinodal curve (black dashed line) on p–ρ plane for intermolecular interaction $\phi_{b5}(r)$ and isotherm (solid line) for critical temperature. Black circle: critical point. Black solid line: isothermal curve obtained from the energy equation. Red solid line: isothermal curve obtained from the compressibility equation. Since the red solid line and the black solid line coincide with each other on the scale in the figure, the red solid line is hidden behind the black solid line and cannot be seen.

necessary to recalculate using the pressure equation derived from the compressibility equation. These matters are explained using the calculation results obtained for the interparticle interaction ϕ_{b5}.

First, find the spinodal curve that satisfies $A(\rho, \beta) = 0$ using the energy equation. The result is shown by a broken line in the p–ρ diagram of Fig. 7.1. In this figure, black filled circle indicates critical point. The black and red solid lines show the p–ρ curves for the critical temperatures obtained from the energy equation and the compressibility equation, respectively. However, since they are the same on the scale in the figure, the red solid line is hidden behind the black solid line and cannot be seen. From this, it can be seen that the pressures derived based on the SCOZA theory, from the two different paths of the energy equation and the compressibility equation agree with high accuracy. However, strictly speaking, there is a minute mismatch between the two. In order to clarify this, an enlarged map near the critical point is shown in Fig. 7.2 for $\beta = \beta_c$ ($= 1.11664$). An enlarged view of the p–ρ curve for $\beta = 1.12041$

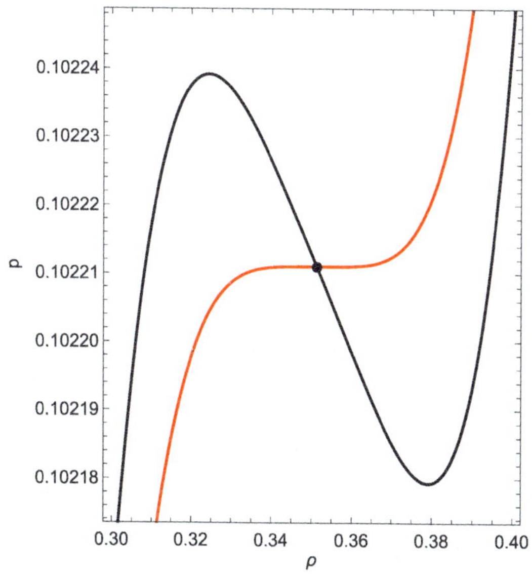

Figure 7.2 Isotherm ($\beta_c = 1.11664$) for intermolecular interaction $\phi_{b5}(r)$. Black circle: critical point. Black solid line: p–ρ curve for the critical temperature obtained from the energy equation. Red solid line: p–ρ curve for the critical temperature obtained from the compressibility equation.

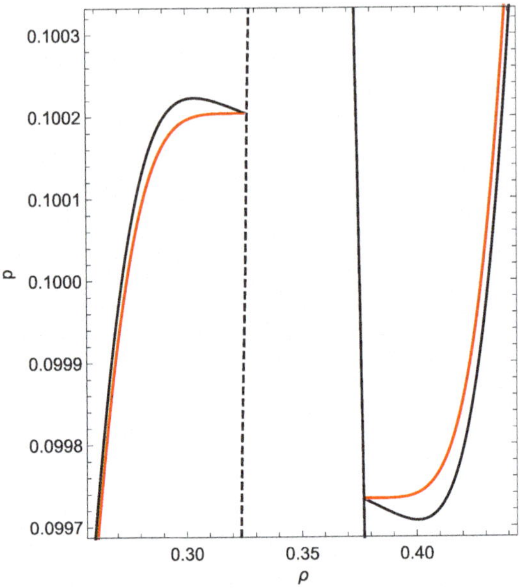

Figure 7.3 Isotherms ($\beta = 1.12041$) and spinodal curves for intermolecular interaction $\phi_{b5}(r)$. Black solid lines: p–ρ curve obtained from the energy equation. Red solid lines: p–ρ curve obtained from the compressibility equation. Dashed lines: spinodal lines.

is also shown in Fig. 7.3. In this enlarged view, it can be seen that there is a slight deviation between the two. The pressure (black line) obtained from the energy equation does not satisfy $A(\rho, \beta) = 0$, that is, $(\partial p/\partial \rho)_\beta = 0$ at the critical point. The points that satisfy $A(\rho, \beta) = 0$ appear in two places on the low-density side and the high-density side that are a little away from the critical point. On the other hand, of course, such a contradiction does not occur in the pressure (red line) obtained from the compressibility equation.

Therefore, in order to obtain the pressure consistently, calculate the pressure on the spinodal curve using the formula obtained from the energy equation. Using the result as the initial value, it is necessary to recalculate the pressure using the formula obtained from the compressibility equation. In this way, the above contradiction is resolved. From now on, pressure will be calculated according to such a procedure. As will be described later, the same operation is performed for the chemical potential.

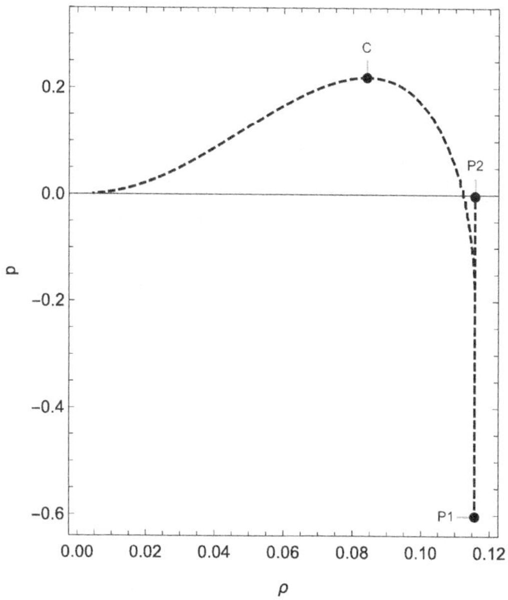

Figure 7.4 Spinodal curve on p–ρ plane for intermolecular interaction $\phi_{c1}(r)$.

Revert story to ϕ_{c1}. The p–ρ diagram of Fig. 7.4 shows the spinodal curve with a broken line obtained using the energy equation. In addition, a part of the p–ρ diagram in Fig. 7.5 is indicated by a broken line. In the figures, the point C ($\rho_c = 0.0841$, $p_c = 0.21935$) represents the critical point. On the lower density side than the critical point C, the pressure p on the spinodal curve increases monotonously with the increase of the density ρ. On the other hand, on the higher density side than the critical point C, the pressure on the spinodal curve decreases as the density increases, reaching a minimum at the point P_1 ($\rho = 0.1157$, $p = -0.6021$). After that, on the spinodal curve, the pressure increases and reaches the point P_2 ($\rho = 0.1157$, $p = -0.002525$) while keeping the density constant.

7.3.1.1 Isotherms

When $T < T_c$, the pressure must be calculated separately for the gas region and the liquid region under the boundary condition

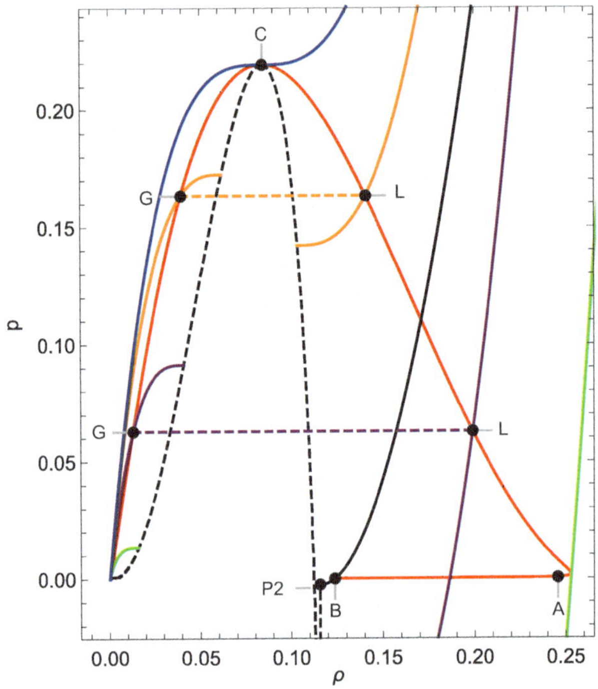

Figure 7.5 Isotherms and gas–liquid coexistence region. The blue, orange, purple, green, and black lines show the isotherms for $\beta = \beta_c = 0.107891$, 0.1154, 0.1414, 0.2684, and 0.5954, respectively. Black circle C is a critical point. The black circle P_2 represents the intersection of the spinodal curve and the black solid line. The red solid line and the black broken line represent the binodal curve and the spinodal curve, respectively.

set on the spinodal curve. Using $\beta p(\rho, \beta)$ on the spinodal curve obtained from the energy equation as the initial value, you can get $\beta p(\rho_j, \beta_i)$ for any step below the critical temperature from the pressure equation obtained from the compressibility equation according to the methods explained in Sections 7.1 and 7.3.1.

The isotherms of pressure obtained in this way are shown in Fig. 7.5 for $\beta = \beta_c$ (= 0.107891), 0.1154, 0.1414, 0.2684, and 0.5954, respectively, by blue, orange, purple, green, and black curves. These temperatures are the same as the representative ones mentioned in Section 6.2.4. The black circle C in the figure represents the critical point. The black dashed line represents the spinodal curve. The solution of Eq. (5.124) is unstable in the region inside the spinodal curve. Of note is the behavior of the isotherm of the liquid. The isotherm moves to the low pressure side until the temperature drops to reach $\beta = 0.2684$ (green line), just like normal substances. However, when it is further cooled below this temperature, the isotherm reverses and moves to the high pressure side, and behaves in the opposite manner to a normal substance. That is, the low temperature black line is located on the higher pressure side than the higher temperature purple line and green line. This strange behavior produces negative thermal expansion, as explained in Section 2.2.

In the gas region, the isotherm monotonously moves to the low pressure side as the temperature decreases.

7.4 Chemical Potential Derived from the Energy Equation

Integrating both sides of the energy equation (1.20) with respect to β in the range between β_* ($\ll \beta_1$) and β_i ($i \geq 1$), we get the following equation:

$$\beta\mu(\rho, \beta_i) = \beta\mu(\rho, \beta_*) - \frac{3}{2}\ln\beta_* + \frac{3}{2}\ln\beta_i + \int_{\beta_*}^{\beta_i}\left(\frac{\partial u}{\partial \rho}\right)_\beta d\beta. \quad (7.2)$$

7.5 Chemical Potential at $\beta = \beta_*$

If $\beta_* \ll 1$, the thermal kinetic energy per molecule is much larger than the interparticle interaction $\phi_{c1}(r)$, so the fluid can be considered as an ideal gas. In this case, the first term and the second term on the right-hand side of Eq. (7.2) can be obtained as follows:

Consider an ideal gas consisting of $\mathcal{N}$ monatomic molecules with a mass of m. The Hamiltonian of this system is

$$\mathscr{H} = \sum_{i=1}^{3\mathcal{N}} \frac{p_i^2}{2m}.$$

Here, an orthogonal coordinate system was adopted and serial numbers were assigned to the coordinate components. Considering the canonical distribution, the partition function is

$$Z = \frac{1}{h^{3\mathcal{N}}\mathcal{N}!}\int e^{-\beta\mathscr{H}}dq_1\cdots dq_{3\mathcal{N}}dp_1\cdots dp_{3\mathcal{N}}$$
$$= \frac{V^{\mathcal{N}}}{h^{3\mathcal{N}}\mathcal{N}!}\left(\int_{-\infty}^{\infty} e^{-\beta p^2/2m}dp\right)^{3\mathcal{N}} = \frac{V^{\mathcal{N}}}{h^{3\mathcal{N}}\mathcal{N}!}\left(\frac{2\pi m}{\beta}\right)^{3\mathcal{N}/2}. \quad (7.3)$$

In this case, Helmholtz's free energy is calculated using Stirling's formula $\ln\mathcal{N}! \approx \mathcal{N}\ln\mathcal{N} - \mathcal{N}$,

$$F = -kT\ln Z$$
$$= -kT\left[\mathcal{N}\ln V + \frac{3\mathcal{N}}{2}\ln(2\pi mkT) - 3\mathcal{N}\ln h - \mathcal{N}\ln\mathcal{N} + \mathcal{N}\right]. \quad (7.4)$$

From this, the chemical potential of the ideal gas is

$$\beta\mu^{\mathrm{id}} = \beta\left(\frac{\partial F}{\partial\mathcal{N}}\right)_{T,V} = -\ln V - \frac{3}{2}\ln(2\pi mkT) + 3\ln h + \ln\mathcal{N}$$
$$= \ln\rho + \frac{3}{2}\ln\beta + \ln\left(\frac{h^2}{2\pi m}\right)^{3/2} = \ln\left(\rho\Lambda^3\right). \quad (7.5)$$

Here, Λ is the de Broglie thermal wavelength and is defined by the following equation:

$$\Lambda = \left(\frac{h^2}{2\pi mkT}\right)^{1/2}. \quad (7.6)$$

The chemical potential obtained by Carnahan–Staring for the hard sphere fluid (see Fig. 7.6) is given by

$$\beta\mu_{\mathrm{CS}} = \ln\rho + \frac{8\eta - 9\eta^2 + 3\eta^3}{\Delta^3}. \quad (7.7)$$

In this book, we replace $\ln\rho$ in Eq. (7.5) with $\beta\mu_{\mathrm{CS}}$, and propose the following equation as a chemical potential for a high-temperature gas:

$$\beta\mu(\rho,\beta_*) = \beta\mu_{\mathrm{CS}} + \ln\left(\frac{h^2\beta_*}{2\pi m}\right)^{3/2}, \quad (\beta_* \ll 1). \quad (7.8)$$

Substituting Eq. (7.8) into Eq. (7.2) gives the following expression:

$$\beta\mu(\rho,\beta_i) = \beta\mu_{\mathrm{CS}} + \ln\left(\frac{h^2\beta_i}{2\pi m}\right)^{3/2} + \int_{\beta_*}^{\beta_i}\left(\frac{\partial u}{\partial\rho}\right)_{\beta} d\beta. \quad (7.9)$$

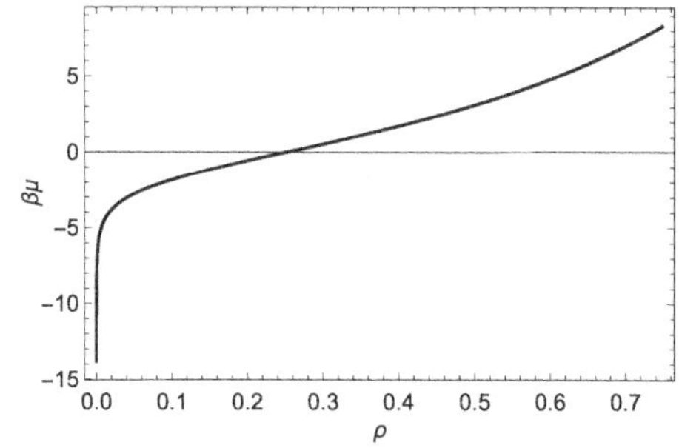

Figure 7.6 Carnahan–Staring's chemical potential $\beta\mu_{\text{CS}}$ for hard sphere fluid.

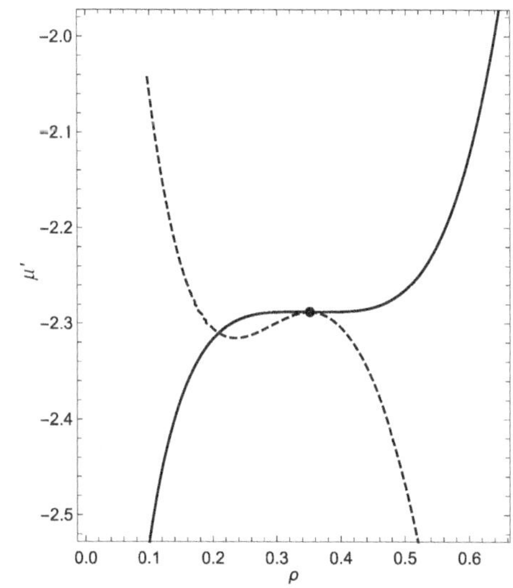

Figure 7.7 The same as Fig. 7.1 except for μ'.

The second term on the right-hand side depends on temperature but not on density, so it has no effect on the determination of liquefaction and vaporization points (see Section 8.1). So we introduce the next μ'.

$$\begin{aligned}\beta\mu'(\rho, \beta_i) &= \beta\mu(\rho, \beta_i) - \ln\left(\frac{h^2\beta_i}{2\pi m}\right)^{3/2} \\ &= \beta\mu_{\text{CS}} + \int_{\beta_*}^{\beta_i}\left(\frac{\partial u}{\partial\rho}\right)_\beta d\beta. \qquad (7.10)\end{aligned}$$

Using the trapezoidal formula for integration, the difference of $\beta\mu'(\rho, \beta_{i+1})$ and $\beta\mu'(\rho, \beta_i)$ yields the following:

$$\beta\mu'(\rho, \beta_{i+1}) = \beta\mu'(\rho, \beta_i) + \frac{\Delta\beta}{2}\left[\left(\frac{\partial u}{\partial\rho}\right)_{\beta_i} + \left(\frac{\partial u}{\partial\rho}\right)_{\beta_{i+1}}\right], \quad (7.11)$$

where,

$$\beta\mu'(\rho, \beta_* = 0) = \beta\mu_{\text{CS}}. \quad (7.12)$$

7.6 Chemical Potential Derived from Compressibility Equation

From Eqs. (1.23) and (5.122), we get

$$\left(\frac{\partial\beta\mu'}{\partial\rho}\right)_\beta = \frac{1}{\rho}\left[\frac{A(\rho, \beta)}{2\pi}\right]^2. \quad (7.13)$$

If the above equation is integrated in both ranges from ρ_1 to ρ_j and ρ_{j+1} with respect to ρ, the following equations are obtained:

$$\beta\mu'(\rho_j, \beta) = \beta\mu'(\rho_1, \beta) + \int_{\rho_1}^{\rho_j}\frac{1}{\rho}\left[\frac{A(\rho, \beta)}{2\pi}\right]^2 d\rho.$$

$$\beta\mu'(\rho_{j+1}, \beta) = \beta\mu'(\rho_1, \beta) + \int_{\rho_1}^{\rho_{j+1}}\frac{1}{\rho}\left[\frac{A(\rho, \beta)}{2\pi}\right]^2 d\rho.$$

From the above two equations and the trapezoidal formula for integration, the following equation is obtained:

$$\beta\mu'(\rho_{j+1}, \beta) = \beta\mu'(\rho_j, \beta) + \frac{\rho_{j+1} - \rho_j}{8\pi^2}\left[\frac{A(\rho_j, \beta)^2}{\rho_j} + \frac{A(\rho_{j+1}, \beta)^2}{\rho_{j+1}}\right]. \quad (7.14)$$

Using μ' derived from the energy equation, find the value on the spinodal curve, use it as an initial value, and use the above equation obtained from the compressibility equation to get μ' at any other point (ρ_j, β_i). The calculation method to obtain μ' is the same as in the case of pressure.

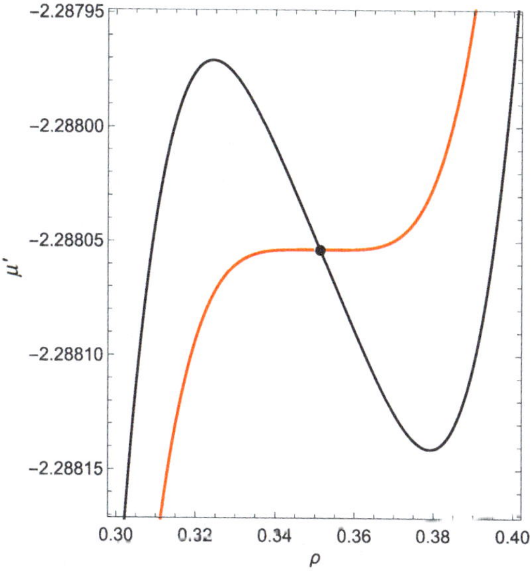

Figure 7.8 The same as Fig. 7.2, except for μ'.

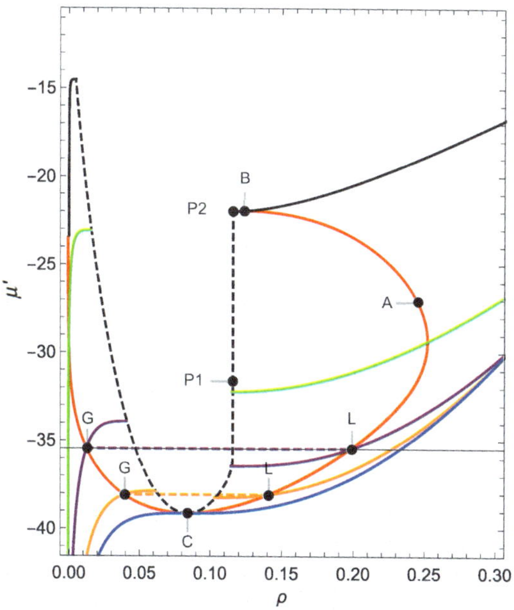

Figure 7.9 The same as Fig. 7.5, but for μ'.

Also, μ' obtained from each of the energy equation and the compressibility equation agrees with high accuracy, but there is a slight inconsistency between the two as in the case of pressure. We confirm these matters by taking the calculation results obtained for ϕ_{b5}. The chemical potential $\mu'(\rho, \beta)$ for the critical temperature obtained from each of the energy equation and compressibility equation is shown, respectively, by the black line and the red line in Fig. 7.7. On the scale of the figure, the slight deviation between the two is negligibly small, so the red solid line is hidden behind the black solid line. Figure 7.8 shows an enlarged view near the critical point. As explained with regard to pressure, the critical point must be $(\partial\mu'/\partial\rho)_\beta = 0$, but the former case is not so and there is a contradiction. However, the latter does not cause such a contradiction. The same is true for other temperatures.

Returning to the case of ϕ_{c1}, the isotherm and spinodal curve of the chemical potential μ' obtained through the above calculation procedure are shown in Fig. 7.9. This figure is the same as Fig. 7.5 except that the pressure p is replaced by μ'.

Chapter 8

Thermodynamic Properties of Subcritical Fluids

To better understand the thermodynamic behavior of fluids below the critical temperature, consider the pressure isotherm for $\beta = 0.1154$ in the case that the interparticle interaction is ϕ_{c1} (see orange line in Fig. 7.5). As the density increases in the gas region, the pressure increases along the orange solid line on the low-density side. If you carefully experiment with all the small perturbations off, the pressure will increase along the orange solid line until you reach the spinodal curve (black dashed line). However, if an experiment with some minute perturbation is performed, after the intersection G (liquefaction point: see the next section) of the orange solid line and the red solid line is reached, the pressure becomes constant and proceeds along the horizontal dashed orange line as the density increases. Then, the gas changes into a liquid while the gas and the liquid coexist. When all of the gas turns into a liquid, you reach the intersection L (vaporization point: see the next section) of the red solid line and orange solid line on the right side. After that, the pressure increases rapidly with increasing density along the orange solid line on the right side. The region from the liquefaction point G to the spinodal line is a metastable region, and depending on the

The Physics of Liquid Water
Makoto Yasutomi

ISBN 978-981-4877-25-1 (Hardcover), 978-1-003-05616-4 (eBook)
www.jennystanford.com

presence or absence of perturbation, it is determined whether the change in the thermodynamic state proceeds along the orange solid line or the broken line. The same is true for the metastable region between the right spinodal curve and the vaporization point L. A line connecting vaporization points at different temperatures or a line connecting liquefaction points is called a binodal line.

8.1 Method to Find Liquefaction Point G and Vaporization Point L

Similar thermodynamic state changes are also observed for any isotherm in the range $\beta^{\dagger} \leq \beta \leq \beta^{\ddagger}$ (e.g., purple lines: $\beta = 0.1414$ in Fig. 7.5). Here, $\beta^{\dagger} = 0.107897$ and $\beta^{\ddagger} = 0.5954$. The pressure at the vaporization point (point B in Fig. 7.5) at $\beta = \beta^{\ddagger}$ has the minimum positive value ($p = 0.00001257$) in this computation. Therefore, a binodal curve cannot be drawn for temperatures below $\beta^{\ddagger}$.

The difference between $\beta^{\dagger}$ and β_c is very small (0.000006), so it is unclear whether it can be verified experimentally. However, in this calculation, it was found that there are no vaporization point and liquefaction point in the temperature range of $\beta_c < \beta < \beta^{\dagger}$. It is too narrow to display the absent area of the liquefaction and vaporization points in Figs. 7.5 and 7.9. Instead, it is shown in the enlarged view near the critical point exibited in Figs. 8.1, 8.2, and 8.3. In fact, many different researchers have so far conducted numerous numerical experiments and reported that the temperature range where the liquefaction point and vaporization point cannot be determined exists in a narrow temperature range just below the critical temperature [Rio (2002); Duda (2009); Elliott and Hu (1999); Liu (2005); Lopez (2006); Pagan and Gunton (2005); Skibinsky (2004); Vega (2009, 1992)]. The reason was a long-standing mystery, but this is elucidated in detail in Section 8.2.

Liquefaction point and vaporization point for any β in the temperature range $\beta^{\dagger} \leq \beta \leq \beta^{\ddagger}$ (see Table 8.1) can be obtained from the following equations:

$$p(\rho_G, \beta) = p(\rho_L, \beta), \quad \mu'(\rho_G, \beta) = \mu'(\rho_L, \beta). \tag{8.1}$$

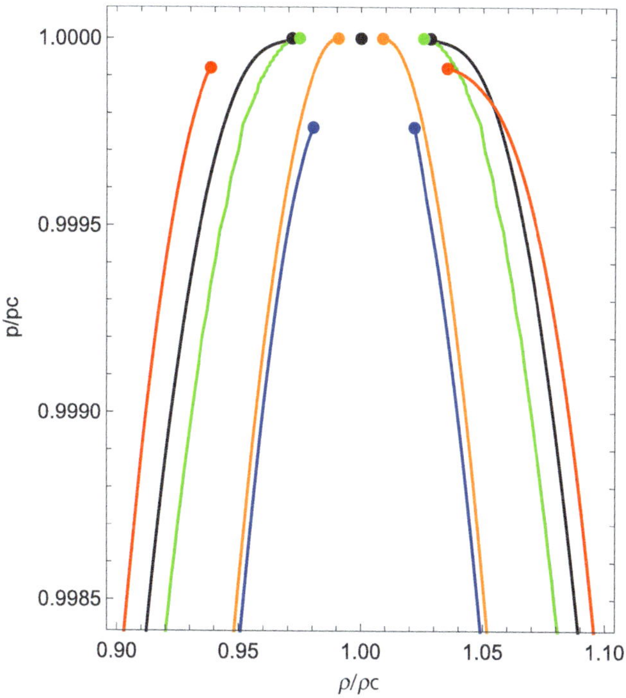

Figure 8.1 The binodal curve absent region on the p–ρ plane for some interparticle interactions. Point C: Critical point. Solid curves of black, orange, green, red, and blue show the binodal curves, respectively, for ϕ_{b2}, ϕ_{c3}, ϕ_{b3}, ϕ_{b5}, and ϕ_{c1}. There is no liquefaction point and vaporization point between each color filled circle and the critical point. Each pressure and density was normalized by the value at the corresponding critical point.

In other words, on the pressure isotherms, the pressure at the liquefaction point density ρ_G is equal to that at the vaporization point density ρ_L. And at the same time, on the isotherms of chemical potential, the chemical potential at ρ_G is equal to that at ρ_L.

For different β in the temperature range $\beta^\dagger \le \beta \le \beta^\ddagger$ where liquefaction and vaporization points exist, the liquefaction points (p_G, ρ_G) and the vaporization points (p_L, ρ_L) are shown as red solid lines in Fig. 7.5. What should be noted here is that the density of the binodal curve on the liquid side is maximum when $\beta = 0.2684$ (the intersection of the green and red lines). That is, the density decreases both when the temperature is raised and lowered. In

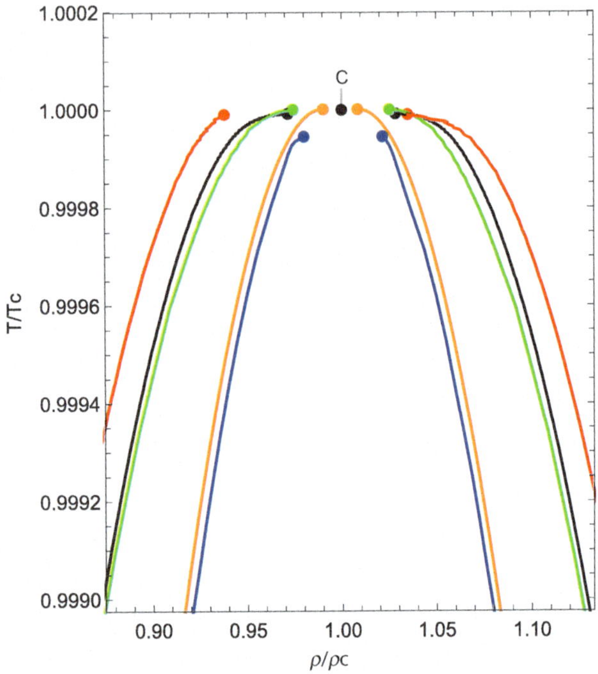

Figure 8.2 The same as Fig. 8.1, but for the $T-\rho$ plane.

Table 8.1 Temperature, density, pressure, and chemical potential at critical points and $\beta^\dagger$ and $\beta^\ddagger$ for some intermolecular interactions

	ϕ_{c1}	ϕ_{c3}	ϕ_{b2}	ϕ_{b3}	ϕ_{b5}
β_c	0.107891	0.145178394	0.810477	0.792344698	1.116636
ρ_c	0.0841	0.2163	0.2463	0.296	0.3513
p_c	0.219349	0.529067	0.0971723	0.121545	0.102211
$\lvert\mu_c'\rvert$	39.1254	20.288406	3.60371	3.43821	2.28805
$\beta^\dagger$	0.107897	0.145178406	0.810483	0.792344706	1.116648
$\beta^\ddagger$	0.5954	0.6869561			

addition, the positive and negative switching of thermal expansion occurs around this temperature.

Here, we explain how to determine the liquefaction point G and the vaporization point L by using the orange isotherms in Figs. 7.5 and 7.9 ($\beta = 0.1154$) as an example. In Fig. 8.4, the isotherms of

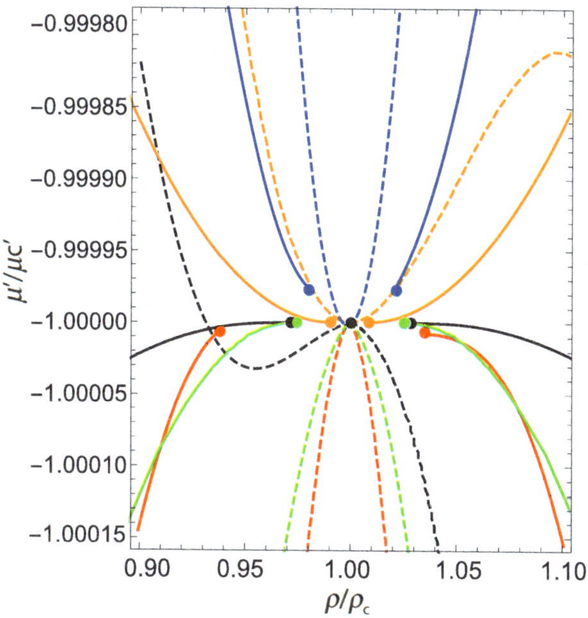

Figure 8.3 The same as Fig. 8.1 except for the μ'–ρ plane. The spinodal curves (dashed lines) are also shown by the same colors as those in Fig. 8.1.

the pressure near the liquefaction point and the vaporization point are indicated by blue broken lines (partly solid lines). Similarly, the isotherms of the chemical potential μ' are shown in Fig. 8.5 with red broken lines (partly solid lines). On the pressure isotherm of the liquid side, pressure and density are minimal at the intersection point $(\rho_{S_2}, p(\rho_{S_2}))$ of the isotherm and the spinodal line. Also, the density on the isotherm on the gas side that takes the same pressure as $p(\rho_{S_2})$ is $\rho = \rho_1$. On this gas side isotherm, at the intersection $(\rho_{S_1}, p(\rho_{S_1}))$ of the spinodal line and the isotherm, both pressure and density are maximum (see Fig. 8.4). The density of the isotherm on the liquid side having the same pressure as $p(\rho_{S_1})$ is ρ_8. Therefore, ρ_G and ρ_L that satisfy the pressure equation of Eq. (8.1) are in the range of $\rho_1 \le \rho_G \le \rho_{S_1}$ and $\rho_{S_2} \le \rho_L \le \rho_8$.

Also for the chemical potential isotherm (see Fig. 8.5), a similar argument shows that ρ_G and ρ_L which satisfy the chemical potential

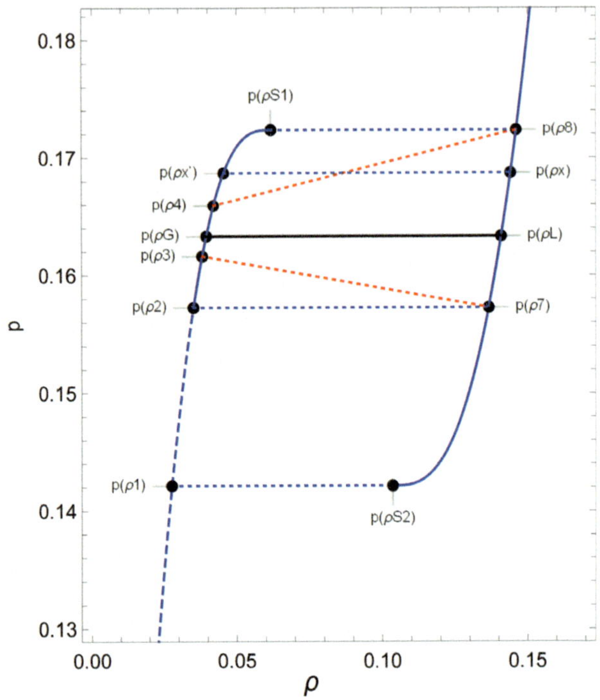

Figure 8.4 Finding vaporization and liquefaction points from pressure isotherms.

equation (8.1) exist in the range given by $\rho_2 \leq \rho_G \leq \rho_{S_1}$ and $\rho_{S_2} \leq \rho_L \leq \rho_9$. Where ρ_2 and ρ_9 are obtained by solving the following two equations $\mu'(\rho_2) = \mu'(\rho_{S_2})$ and $\mu'(\rho_{S_1}) = \mu'(\rho_9)$. Therefore, in order for ρ_G and ρ_L to satisfy the two equations (8.1) at the same time, because $\rho_1 < \rho_2$ and $\rho_8 < \rho_9$, ρ_G and ρ_L should exist in the range of $\rho_2 \leq \rho_G \leq \rho_{S_1}$ (gas side solid line part in Fig. 8.4) and $\rho_{S_2} \leq \rho_L \leq \rho_8$.

Furthermore, the density ρ_7 on the liquid side pressure isotherm that takes the same pressure as the minimum pressure on the gas side isotherm is obtained from $p(\rho_2) = p(\rho_7)$. Therefore, the range of vaporization point density ρ_L is further narrowed to $\rho_7 \leq \rho_L \leq \rho_8$ (see solid line part on the liquid side in Fig. 8.4).

Similarly, considering the chemical potential, $\mu'(\rho_7) = \mu'(\rho_3)$ and $\mu'(\rho_8) = \mu'(\rho_4)$, so the range of ρ_G is narrowed to $\rho_3 \leq \rho_G \leq \rho_4$ (see the gas side solid line part in Fig. 8.5). In this way, if the ranges of

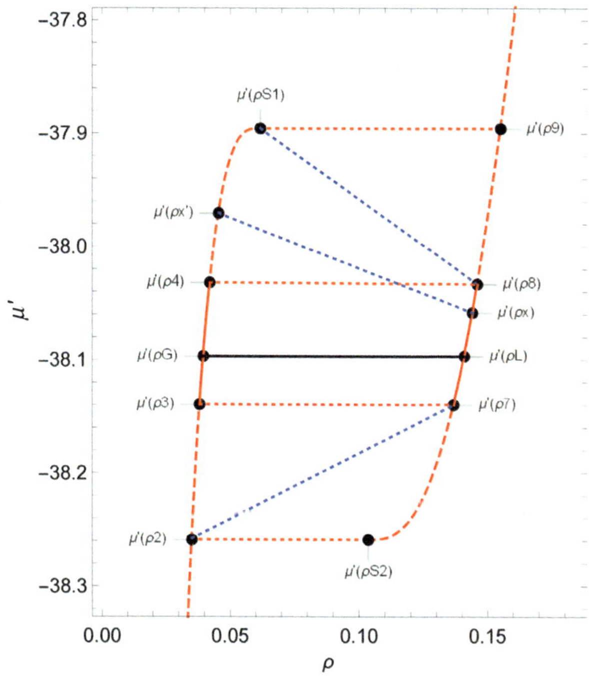

Figure 8.5 Finding vaporization and liquefaction points from chemical potential isotherms.

ρ_{G} and ρ_{L} are narrowed repeatedly, the liquefaction point and the vaporization point can be obtained. However, those two points can be determined in a easier way than that as follows.

As discussed above, $\rho_2 \leq \rho_{\mathrm{G}} \leq \rho_{\mathrm{S_1}}$, $\rho_7 \leq \rho_{\mathrm{L}} \leq \rho_8$, $p(\rho_{\mathrm{S_1}}) = p(\rho_8)$, and $p(\rho_2) = p(\rho_7)$. So, as you can see from the blue solid line in Fig. 8.4, for any density ρ_x on the liquid side blue solid line, only one $\rho_{x'}$ that satisfies $p(\rho_{x'}) = p(\rho_x)$ is determined on the blue solid line on the gas side. Therefore, $\rho_{x'}$ can be obtained numerically as a function of ρ_x which is written as $\rho_{x'} = f(\rho_x)$. Here, if ρ_x changes from ρ_7 to ρ_8, while maintaining $p(\rho_x) = p(\rho_{x'})$, $\rho_{x'}$ varies from ρ_2 to $\rho_{\mathrm{S_1}}$. However, as seen from Fig. 8.5 and Table 8.2, $\mu'(\rho_2) < \mu'(\rho_7)$ and $\mu'(\rho_{\mathrm{S_1}}) > \mu'(\rho_8)$. On the μ' isotherm, μ' is a monotonically increasing function of ρ, there is only one ρ_x between ρ_7 and ρ_8 which satisfies $\Delta\mu'(\rho_x) = \mu'(\rho_{x'}) - \mu'(\rho_x) = 0$. In this case, ρ_x is ρ_{L}, and $\rho_{x'}$ is ρ_{G}. Figure 8.6 shows that $\Delta\mu'(\rho_x)$ is 0 at $\rho_x = \rho_{\mathrm{L}}$.

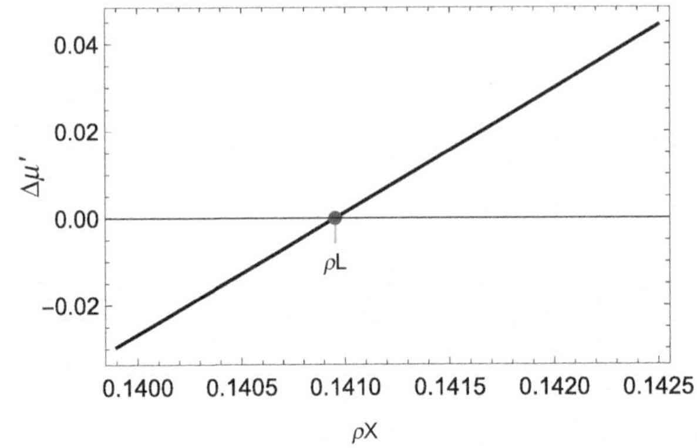

Figure 8.6 $\Delta\mu'(\rho_x) = \mu'(\rho_{x'}) - \mu'(\rho_x)$. $\Delta\mu'(\rho_x) = 0$ gives $\rho_L = \rho_x$ (solid) and $\rho_G = f(\rho_x)$.

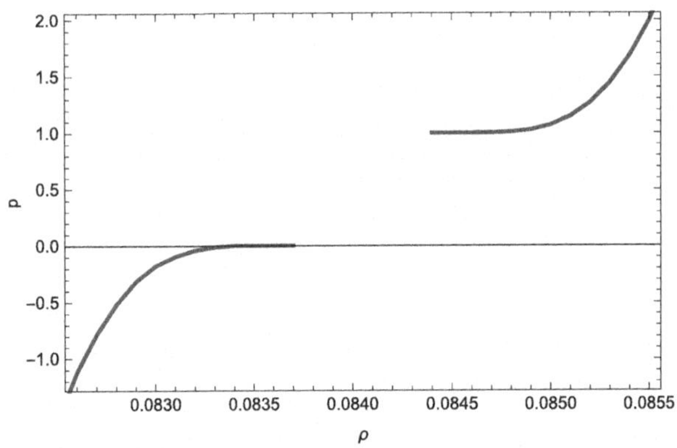

Figure 8.7 The $\langle p\rangle$–ρ curve for $\beta = 0.107894$ just below the critical temperature for ϕ_{c1}.

As shown in Fig. 7.5, ρ_G always decreases monotonically with decreasing temperature. On the other hand, ρ_L increases monotonically in the range $\beta^\dagger < \beta < 0.2684$, but starts to decrease monotonically at $\beta > 0.2684$ with decreasing temperature. And, ρ_L reaches the smallest $\rho_L = 0.1236$ (the pressure is negative below that density) at the intersection point B of the binodal curve and the black isothermal curve ($\beta = \beta^\ddagger$) before the black pressure isotherm reaches the spinodal line.

Table 8.2 Characteristic density, pressure, and chemical potential near the liquefaction and vaporization points for $\beta = 0.1154$

ρ_1	ρ_2	ρ_3	ρ_G	ρ_4	ρ_{S_1}
0.02755	0.03515	0.03815	0.03956	0.04205	0.0618
$p(\rho_1)$	$p(\rho_2)$	$p(\rho_3)$	$p(\rho_G)$	$p(\rho_4)$	$p(\rho_{S_1})$
0.14213	0.15726	0.1616	0.1633	0.1659	0.1724
	$\mu'(\rho_2)$	$\mu'(\rho_3)$	$\mu'(\rho_G)$	$\mu'(\rho_4)$	$\mu'(\rho_{S_1})$
	−38.2591	−38.1396	−38.0969	−38.0319	−37.8958
ρ_{S_2}	ρ_7	ρ_L	ρ_8	ρ_9	
0.1036	0.13665	0.1409	0.1461	0.1551	
$p(\rho_{S_2})$	$p(\rho_7)$	$p(\rho_L)$	$p(\rho_8)$		
0.14213	0.15726	0.1633	0.1724		
$\mu'(\rho_{S_2})$	$\mu'(\rho_7)$	$\mu'(\rho_L)$	$\mu'(\rho_8)$	$\mu'(\rho_9)$	
−38.2591	−38.1398	−38.0969	−38.0332	−37.8958	

8.2 Absence Temperature Region of Liquefaction Point and Vaporization Point ($\beta_c < \beta < \beta^\dagger$)

As mentioned in the introduction of Chapter 8, a number of related numerical simulations have been performed by many different scientists so far [Rio (2002); Duda (2009); Elliott and Hu (1999); Liu (2005); Lopez (2006); Pagan and Gunton (2005); Skibinsky (2004); Vega (2009, 1992)]. However, in a narrow temperature range ($\beta_c < \beta < \beta^\dagger$) just below the critical temperature (see Table 8.1), no one could determine the liquefaction point and vaporization point. The reason was a mystery for many years. However, our study revealed that there is a region without vaporization and liquefaction points just below the critical temperature. This is described in detail below.

As mentioned above, in order for the vaporization point and the liquefaction point to exist, density ρ_{S_i} ($i = 1, 2$) should satisfy the following equations:

$$\Delta p_{21}(\beta) = p(\rho_{S_2}) - p(\rho_{S_1}) \leq 0 \quad \text{and}$$
$$\Delta\mu'_{21}(\beta) = \mu'(\rho_{S_2}) - \mu'(\rho_{S_1}) \leq 0. \tag{8.2}$$

The pressure isotherm $\langle p\rangle$ is shown in Fig. 8.7 for ϕ_{c1} and for $\beta = 0.107894$ in the range of $\beta_c < \beta < \beta^\dagger$. In the figure, $\langle p\rangle$ is defined by

$$\langle p\rangle = \frac{p(\rho, \beta) - p(\rho_{S_1})}{\Delta p_{21}}, \quad \text{where} \quad \Delta p_{21} = 2.27 \times 10^{-7}.$$

As shown in the figure, $\Delta p_{21}(\beta) > 0$. Therefore, it can be seen that there is no liquefaction point and no vaporization point for this temperature. The same is true for any other temperature within that range. Therefore, it can be seen that there are no liquefaction and vaporization points in the range of $\beta_c < \beta < \beta^\dagger$. In order to clarify the existence of this absent region, the enlarged views near the critical point in the p–ρ plane, T–ρ plane, and the μ'–ρ plane are shown in Figs. 8.1, 8.2, and 8.3, respectively. As shown in these figures, the binodal curves are interrupted at the points indicated by blue filled circles, and there are no liquefaction points and vaporization points between the blue filled circles and the critical point (black filled circle).

For ϕ_{b5}, we also show that the absence area of vaporization point and the liquefaction point exist just below the critical point. For this interaction, Eq. (8.2) always holds for temperatures below the critical temperature. However, ρ_G and ρ_L which satisfy simultaneously the two Eqs. (8.1) don't exist in the temperature range of $\beta_c < \beta < \beta^\dagger$. To clarify it, $\Delta\mu'(\rho_x)$ is shown in Fig. 8.8. In the figure, the isothermal curve $\Delta\mu'(\rho_x)$ moves from the top to the bottom as the temperature decreases. The solid circles in the figure represent points that satisfy $\Delta\mu'(\rho_L) = 0$. From this figure, it can be seen that there are no liquefaction point and vaporization point in the region of $\beta_c < \beta < \beta^\dagger$ $(= 1.116648)$. In order to clarify the existence of the absence area of vaporization and liquefaction points also in the cases of the interparticle interactions of ϕ_{b2}, ϕ_{c3}, ϕ_{b3}, and ϕ_{b5}, enlarged views near the critical point are shown in the p–ρ plane Fig. 8.1, T–ρ plane Fig. 8.2, and μ'–ρ plane Fig. 8.3, respectively, by black, orange, green, and red curves. In these figures, pressure, temperature, chemical potential, and density are normalized by the corresponding physical quantities (p_c, T_c, μ'_c, ρ_c) at the critical points obtained for each intermolecular interaction (see Table 8.1).

8.3 Necessity of Two Density Step Sizes

For numerical calculations related to $\phi_{c1}(r)$ and $\phi_{c3}(r)$, two different density step sizes $\Delta\rho_a = 10^{-6}$ and $\Delta\rho_b = 10^{-4}$ are used. The reason

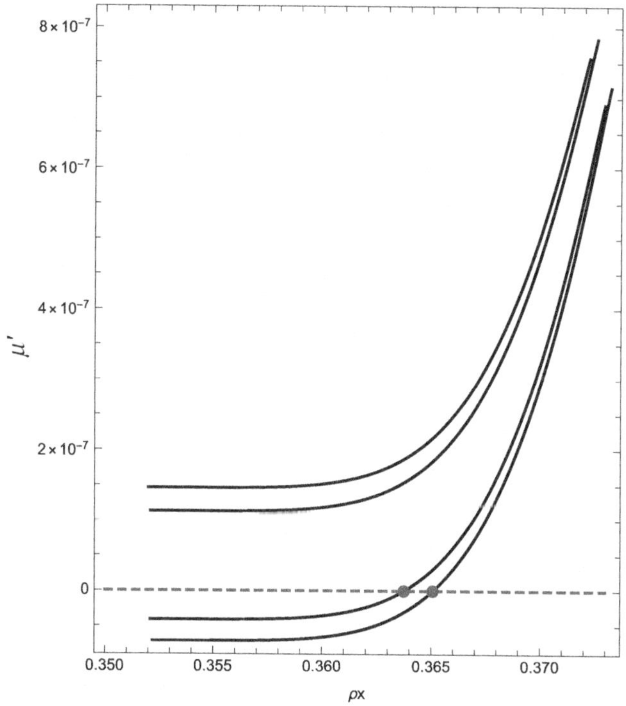

Figure 8.8 $\Delta\mu'(\rho_x)$ for interparticle interaction ϕ_{b5}. The isothermal curve $\Delta\mu'(\rho_x)$ corresponds to $\beta = 1.1166463$, 1.1166473, 1.1166483, and 1.1166493 in order from the top to the bottom of the figure. The solid circles in the figure represent the point where $\Delta\mu'(\rho_L) = 0$.

is described here. The density of vaporization point (black filled circle A) on the liquid side binodal line in Fig. 7.5 obtained for $\phi_{c1}(r)$ is $\rho_L = 0.2463$. The corresponding liquefaction point density is $\rho_G = 10^{-4} = \Delta\rho_b$. Therefore, if the density is engraved with only one step size $\Delta\rho_b$, the resulting binodal line has a limit point A. In order to obtain a binodal line from black filled circle A to black filled circle B, a finer step size $\Delta\rho_a$ than $\Delta\rho_b$ is required in the region of $\rho < 10^{-4}$. Incidentally, the density of the liquefaction point is $\rho_G = 2.349 \times 10^{-6}$. The binodal curve and spinodal curve for $\phi_{c3}(r)$ are shown in Figs. 8.9 and 8.10. In these figures, density ρ_A, pressure p_A, chemical potential μ'_A for point A are $(\rho_A/\rho_c, p_A/p_c, \mu'_A/\mu'_c) = (1.68015, 0.0000621193, -0.775606)$.

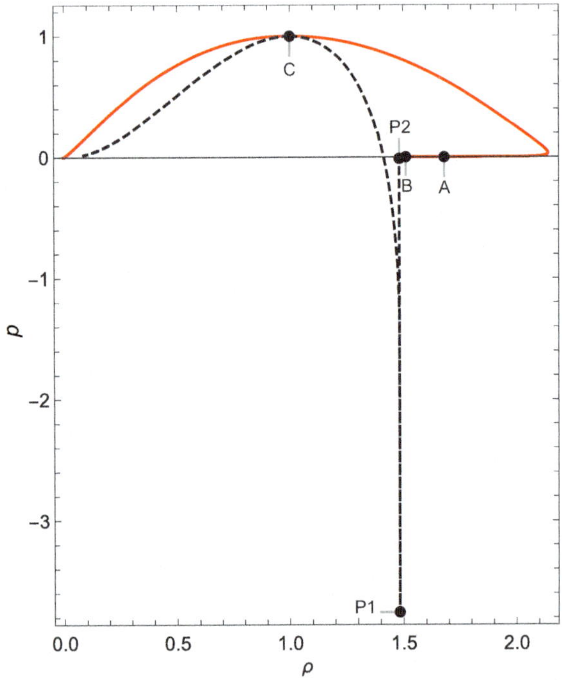

Figure 8.9 Binodal curve (red solid line) and spinodal curve (black dashed line) on p–ρ plane for ϕ_{c3}. Black filled circles A, B, C, P1, and P2 have the same meaning as the black filled circles shown in Fig. 7.5 and Fig. 7.4, respectively. The units of pressure and density are p_c and ρ_c, respectively.

8.4 Comparison of Theory and Experiments

8.4.1 Units of Length σ_u and Energy ε_u

So far, the diameter of the hard sphere is chosen as the unit of length σ_u, and the maximum depth of the intermolecular interaction $\phi(r)$ has been treated as a unit of energy ε_u. The value of these units can be determined by comparing our theoretical results and experimental results. The value of σ_u is determined so that the maximum theoretical density $\rho_{md} \cdot \sigma_u^{-3}$ matches the experimental value of maximum density 999.97 $kg \cdot m^{-3}$ of water under 1 atm. Here, ρ_{md} represents the theoretical calculation value of the maximum density. (Similarly, any other physical quantity f

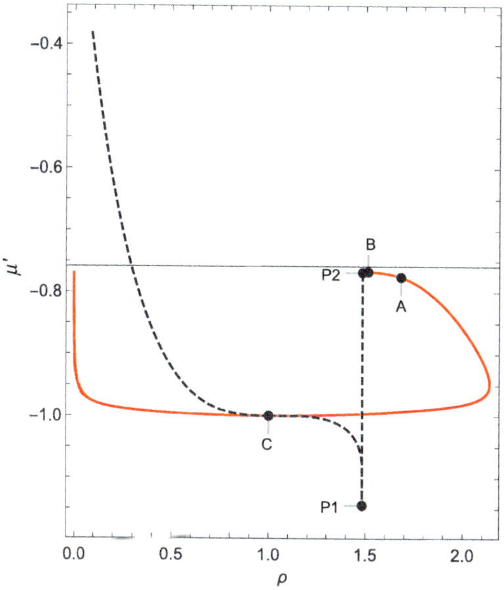

Figure 8.10 The same as Fig. 8.9, but for μ'.

theoretically obtained for this density is represented by f_{md}.)

$$\rho_{\mathrm{md}}\sigma_{\mathrm{u}}^{-3} = \frac{999.97\,\mathrm{kg}\cdot\mathrm{m}^{-3}}{m_{\mathrm{H_2O}}}$$

$$\therefore \quad \sigma_{\mathrm{u}} = 3.10437\times 10^{-10}\rho_{\mathrm{md}}^{1/3}\,\mathrm{m}, \tag{8.3}$$

where, $m_{\mathrm{H_2O}}$ is the mass of the water molecule H_2O.

The unit of energy ε_{u} is decided so that the theoretical value of $kT_{\mathrm{md}}\cdot\varepsilon_{\mathrm{u}}$ matches the experimental value of 3.98°C when water density is maximum at 1 atm.

$$kT_{\mathrm{md}}\cdot\varepsilon_{\mathrm{u}} = (273.15 + 3.98)k\mathrm{K}$$

$$\therefore\ \varepsilon_{\mathrm{u}} = 277.13\beta_{\mathrm{md}}k\mathrm{K} = 3.82622\times 10^{-21}\beta_{\mathrm{md}}\,\mathrm{J}. \tag{8.4}$$

Because 1 atm = 101325 Pa = 101325 J·m^{-3}, the pressure p_{md} is expressed in atmospheric pressure as follows:

$$p_{\mathrm{atm}} = \frac{p_{\mathrm{md}}\cdot\varepsilon_{\mathrm{u}}\cdot\sigma_{\mathrm{u}}^{-3}}{101325\,\mathrm{J}\cdot\mathrm{m}^{-3}}\,\mathrm{atm} = 1262.21\frac{\beta_{\mathrm{md}}\cdot p_{\mathrm{md}}}{\rho_{\mathrm{md}}}\,\mathrm{atm}. \tag{8.5}$$

We have so far calculated isotherms for various temperatures for several interparticle interactions. As shown in the

Table 8.3 Normalization constants p_{md}, ρ_{md}, and β_{md} for the isobaric lines in Fig. 8.11

	Red solid line	Green solid line	Black solid line
p_{md}	0.219349	0.0438698	0.000743712
ρ_{md}	0.27248	0.257034	0.252868
β_{md}	0.271402	0.269402	0.269402

conceptual diagram of Fig. 2.2, the isobaric line for a certain atmospheric pressure is obtained from the set of the intersections of these isothermal lines and the constant pressure line on the p–ρ plane.

From the theoretical isobaric lines for various pressures obtained in this way, the curve to be obtained is the theoretical isobaric line under 1 atm which best reproduces the shape of the isobaric line experimentally obtained under 1 atm and satisfies $p_{atm} = 1$ atm.

As an example, the three isobars obtained for ϕ_{c1} are shown in Fig. 8.11. The red solid line, the green solid line, and the black solid line are theoretical isobaric lines obtained for $p = p_c = 0.219349$ (critical pressure), 0.0438698, and 0.000743712, respectively. The black filled circles represent the isobaric data experimentally obtained under 1 atm, and are the same as the black filled circles shown in Fig. 2.1. Here, the density and temperature of each isobaric line in this figure are normalized by the corresponding ρ_{md} and β_{md} (see Table 8.3). As shown in the figure, the theoretically obtained black and green solid lines reproduce well the blue filled circles obtained from experiments in the range of $-10 \leq T(^{\circ}\mathrm{C}) \leq 100$.

For the green solid line, $\rho_{md} = 0.257034$, $\beta_{md} = 0.269402$, so $\varepsilon_u = 1.0308 \times 10^{-21}$ J, $\sigma_u = 1.974$ Å, $p_{atm} = 58.0373$ barometric pressure. For the black solid line, $\rho_{md} = 0.252868$ and $\beta_{md} = 0.269402$, so $\varepsilon_u = 1.0308 \times 10^{-21}$ J, $\sigma_u = 1.963$ Å, $p_{atm} = 1.0001$ barometric pressure. Therefore, the black solid line is a theoretical isobaric line to be obtained. As shown in Fig. 3.1, the water molecule has a shape close to a sphere with a diameter of about 3 Å, but the size σ_u of the hard sphere obtained from the solid black line is about 2/3 of that of water molecule. So, it makes sense.

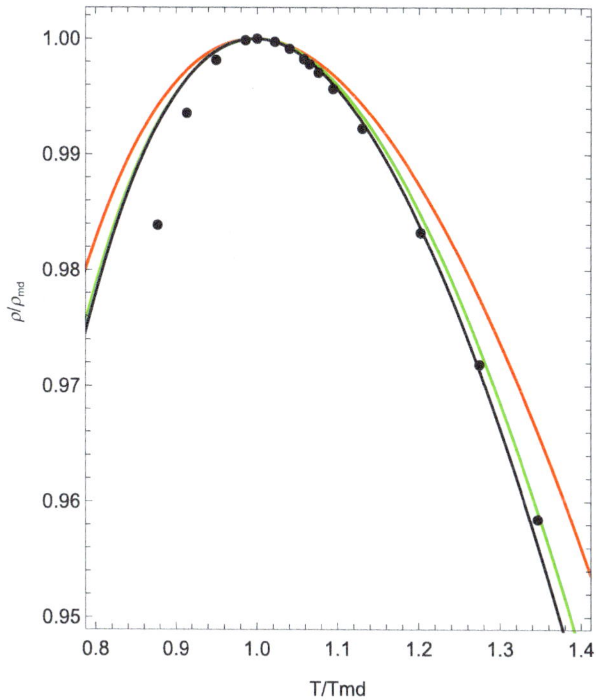

Figure 8.11 Comparison of the theoretically obtained isobaric density-temperature relations and the experimental data. Solid lines: theoretical results; Filled circles: experimental data.

8.4.2 Pressure, Temperature, and Density at Critical Point

Since the theoretical values of pressure, density, and temperature at the critical point obtained for ϕ_{c1} are $p_c = 0.219$, $\rho_c = 0.0841$, and $\beta_c = 0.1079$ (see Table 8.1), using $\varepsilon_u = 1.031 \times 10^{-21}$ J and $\sigma_u = 1.963$ Å obtained for the solid black line in the previous section, the pressure, temperature, and density at the critical point are

$$\frac{p_c \varepsilon_u \sigma_u^{-3}}{101325\,\mathrm{Jm}^{-3}} = 294.9\,\mathrm{atm}, \quad \rho_c \sigma_u^{-3} = 1.112 \times 10^{28}\,\mathrm{m}^{-3},$$
$$\frac{\varepsilon_u}{k\beta_c} = 692\,\mathrm{K}.$$

The experimental values of pressure, volume, and temperature at the critical point are 218.3 atm, $59.1 \pm 0.5\ \mathrm{cm}^3/\mathrm{mol}$, $374.15°\mathrm{C}$

(Table 8.4), the error rates of the theoretical values are

$$\text{Error rate of pressure} = \frac{294.9 - 218.3}{218.3} = 0.35.$$

$$\text{Error rate of temperature} = \frac{692 - 647.3}{647.3} = 0.069.$$

$$\text{Error rate of density} = \frac{1.112 - 1.019}{1.019} = 0.091.$$

Comparing the experimental value and the theoretical value at the critical point, the error rate is the largest in the case of pressure and is 0.35. It is not so good, but not extremely bad.

Table 8.4 Temperature T_c, critical volume V_c, and pressure p_c at the critical point of water [Nowak (1961)]

T_c (Ž)	V_c (cm^3/mol)	p_c (atm)
374.15	59.1±0.5	218.3

8.4.3 Comparison of Theoretical and Experimental Vaporization Points

The measured temperature and density of the water vaporization point at 1 atm are 373.15 K and 0.958 g/cm^3, respectively. On the other hand, the theoretical values determined from the binodal curve (see red line in Fig. 7.5) for the intermolecular interaction ϕ_{c1} are $\beta = 0.3216/\varepsilon_u$ and $\rho = 0.2478/\sigma_u^3$. The optimal unit values obtained from the solid black line shown in Fig. 8.11 are $\varepsilon_u = 1.031 \times 10^{-21}$ J and $\sigma_u = 1.974$ Å, so density and temperature in the units of g/cm^3 and K are represented, as

$$\rho = 0.2478 \cdot \sigma_u^{-3} \cdot m_{H_2O} = \frac{0.2478 \cdot 18\,\text{g}}{(1.974 \cdot 10^{-8}\,\text{cm})^3 \cdot 6.02214 \cdot 10^{23}}$$

$$= 0.9629\,\text{g/cm}^3.$$

$$T = \frac{\varepsilon_u}{k\beta} = \frac{1.031 \cdot 10^{-21}}{0.3216 \cdot 1.3806 \cdot 10^{-23}}\,\text{K} = 232.2\,\text{K}.$$

Therefore, the discrepancies of the theoretical values from the measured values of density and temperature at the liquefaction point are 0.47% and 37.8%, respectively. It can be said that the density is very good. However, the temperature is not so good, but not extremely bad.

8.5 Simplest and Most Optimum Interactions between Water Molecules

So far, it has been assumed that the interaction between water molecules is independent of temperature and density. And, the functional form of the potential tail has been given by multi-Yukawa terms with three or more. Therefore, the potential contains more than 5 parameters, excluding the one normalization constant. It was explained that there are innumerable such interactions between water molecules. But when the parameters are 5 or more, it is extremely difficult to investigate all potentials. In addition, regarding the reproducibility of the experimental results of the ρ–T relation under 1 atmosphere, it is highly accurate in the temperature range from 0°C to +100°C. But, as for the reproducibility for temperatures below 0°C, the gap between the experiment and theory increased with the temperature drop.

Under more realistic conditions, it is considered that the functional shape of the interaction between water molecules changes depending on temperature and density. With that in mind, we reviewed the interactions between water molecules and found the most simple and optimal potential tails given by only two terms of Yukawa potential (one is for soft repulsion and the other for attraction). It was found that the experimental results could be reproduced with extremely high accuracy in the temperature range of −30°C to +100°C. Except for the normalization constant, the potential is given with only three parameters, so it is possible to investigate all functional forms. This finding is very important. As a result, it has become practical to elucidate the mysteries of various anomalous thermal behaviors of liquid water over a wide range of temperature, density, and pressure, by associating them with the functional shape of the potential. I'm sure most of these mysteries can be solved in the coming years.

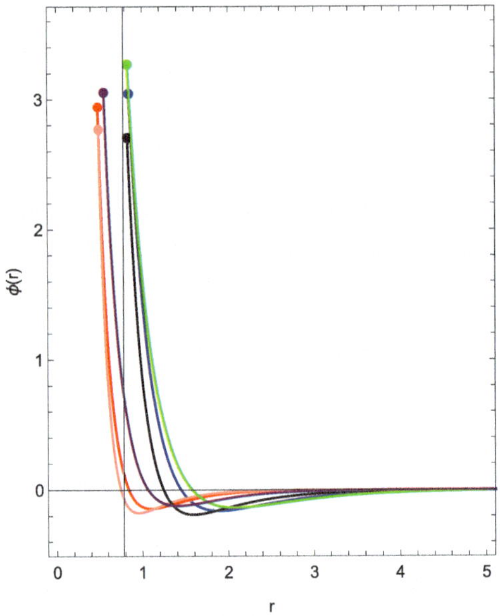

Figure 8.12 The simplest and optimum potential tails of water molecule interactions. The unit of length is Å and the unit of potential tails is 10^{-20} Joule. Blue line: ϕ_A, Red line: ϕ_B, Black line: ϕ_C, Pink line: ϕ_D, Green line: ϕ_E, Purple line: ϕ_F. The circles of each color indicate the value of the tail with same color at the hard-core contact.

The potentials mentioned above are shown in Fig. 8.12. These are given by Eq. (3.2) for $N = 3$. The parameters z_n and a_n ($n = 2, 3$) are shown in Table 8.5.

In Fig. 8.12, the three tails on the right side represent the potential tails in the temperature range of 0 degrees Celsius to +100

Table 8.5 Parameters for ϕ_A, ϕ_B, ϕ_C, ϕ_D, ϕ_E, and ϕ_F

	ϕ_A			ϕ_B			ϕ_C	
n	z_n	a_n	n	z_n	a_n	n	z_n	a_n
2	1	18.6276	2	1	19.1636	2	1	8.42487
3	2	$-2.02\times a_2$	3	2	$-2.06\times a_2$	3	3	$-2.696\times a_2$
	ϕ_D			ϕ_E			ϕ_F	
n	z_n	a_n	n	z_n	a_n	n	z_n	a_n
2	1	8.67064	2	1	42.7371	2	1	43.5842
3	3	$-2.8\times a_2$	3	1.5	$-1.562\times a_2$	3	1.5	$-1.573\times a_2$

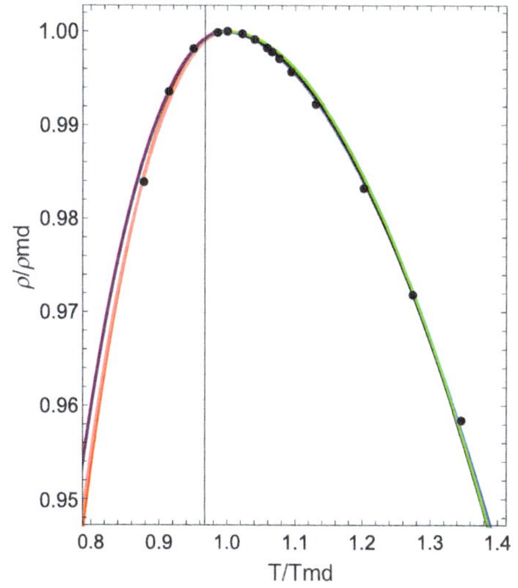

Figure 8.13 Comparison of theoretical and experimental results of ρ–T curve at 1 atm. Black circles: Experimental results. Solid line: theoretical results. The solid line color represents the ρ–T curve obtained from the same color potential shown in the Fig. 8.12. In this figure, the three lines of the black line, the blue line and the green line are so close that they are hard to distinguish. The pink line and the red line are also the same.

degrees Celsius, and the three tails on the left side represent the tails below +10 degrees Celsius.

Also, the ρ–T curve at 1 atm obtained from each potential is displayed in Fig. 8.13 by the same color line as that in Fig. 8.12. As can be seen from the figure, the ρ–T curves obtained from these potentials accurately reproduce the experimental results (black circles) in the temperature range of −30 to +100 degrees Celsius.

Any combination of a potential at higher temperatures and that at lower temperatures is possible. Therefore, in the present case, 9 combinations are possible. From these combinations, it is necessary to calculate isothermal compressibility and other physical quantities and compare them with experimental results to search for a more optimal potential.

Future studies will clarify whether the change between the potential at higher temperatures and that at lower temperatures suddenly occurs, occurs in multiple stages, or occurs continuously.

When the change suddenly occurs, a dramatic change (for example, phase change) may occur in some physical quantity. I look forward to future developments.

8.6 Concluding Remarks

This book has fundamentally elucidated the mystery of the negative thermal expansion of liquid water that could not be solved for centuries. The fundamental elucidation here means that reproducing the phenomenon using thermodynamic quantities derived using the laws of thermodynamics and statistical mechanics, and means that elucidating the cause of the negative thermal expansion in relation to the functional form of the interaction between particles.

As shown in Fig. 2.2, the density–temperature curve under constant pressure is given by the intersections of the pressure isotherms and the line of constant pressure. The compressibility–temperature curve under constant pressure is given by the slope of the pressure isotherm at the intersections mentioned above as seen from Eq. (1.21). The shape of this curve obtained from the experiment is also full of mysteries. We found many interparticle interactions that reproduce the former curve with high accuracy. Among such interactions, it is necessary to select those that simultaneously satisfy the latter curve. Our experience suggests that such interactions are constantly changing depending on the state of thermal equilibrium. I strongly believe that if we proceed with research considering such things, not only these two curves but also the strange thermal behavior of many other thermodynamic quantities can be elucidated in the coming years. Those results will be reported in the next revision of this book. What made this possible was the discovery of the simplest and most optimal water molecule interactions. The interaction consists of two tails represented by Yukawa terms, one giving a soft repulsive force and the other an attractive force. This finding should play an important role in future in studying the thermodynamic properties of water.

According to the theoretical calculation in this book, under one atmosphere, the density of water below 4°C decreases monotonically with decreasing temperature, but according to the experiment [Mallamace (2007)], the density has been reported to have a minimum value of $\rho = 0.924\mathrm{g/cm}^3$ at $T = 203$ K, and then increases slightly. The minimum value of this density is between the density of point A and the density of point B shown in Fig. 7.5. I am convinced that such behavior of the density–temperature curve under constant pressure can be clarified by the method described above. I am confident that this will also be reported in the next revision.

Theoretical calculations in this book and numerical experiments [Rio (2002); Duda (2009); Elliott and Hu (1999); Liu (2005); Lopez (2006); Pagan and Gunton (2005); Skibinsky (2004); Vega (2009, 1992)] show that there are no liquefaction point and vaporization point in the narrow temperature region just below the critical temperature. However, it is not certain at this time whether this is true or depends on the accuracy of the numerical calculations, but the fact is sure that more careful calculations are needed to confirm that.

I used SCOZA in this book, and explained the necessary equations, analytical solutions, and numerical calculation methods so that one can understand them without filling the space between lines. Therefore, I believe that the readers can understand them in a short time and reach a point where they can apply them to specific problems. The purpose of this book will be achieved if many researchers would be able to use the knowledge gained in this book to solve many mysteries about liquids.

Appendix A

Integration Region

A.1 Integration Region 1

The integration region of the first term on the right side of Eq. (5.69) does not exist in the range of $0 \leq s \leq 1/2$. For the range $1/2 < s \leq 1$, the integration region in the t–r plane can be represented by a triangle surrounded by three boundaries $t = r + s$, $t = 1$, and $r = s$. Here, as examples, in the cases of $s = 3/4$ and $s = 1$, the integration regions are shown as green triangles in Fig. A.1 and Fig. A.2, respectively.

A.2 Integration Region 2

The integration region of the second term on the right side of Eq. (5.69) is considered separately for the cases of $0 < s \leq 1/2$ and $1/2 < s < 1$. In the former case, the integration region in the t–r plane can be represented by a triangle surrounded by three boundary lines $t = r + s$, $t = 1$, and $r = 1$. Here, as examples, in the cases of $s = 1/4$ and $s = 1/2$, the integration regions are shown in green in Fig. A.3 and Fig. A.4, respectively.

The Physics of Liquid Water
Makoto Yasutomi

ISBN 978-981-4877-25-1 (Hardcover), 978-1-003-05616-4 (eBook)
www.jennystanford.com

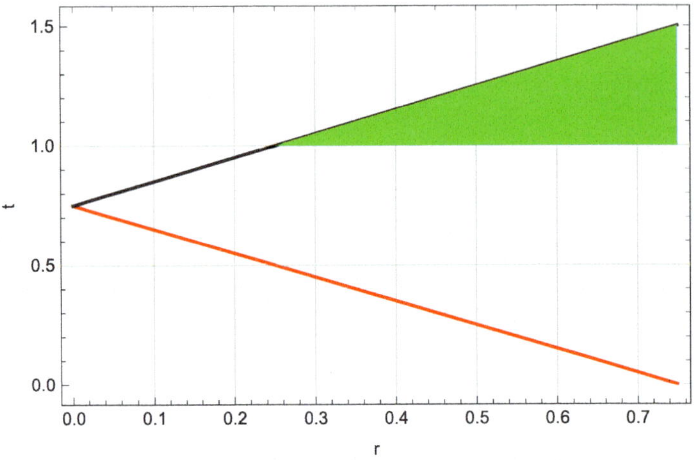

Figure A.1 Integration region (green triangle) in the t–r plane for $s = 3/4$ in the first term on the right-hand side of Eq. (5.69). Black line: $t = r + s$, red line: $t = r - s$.

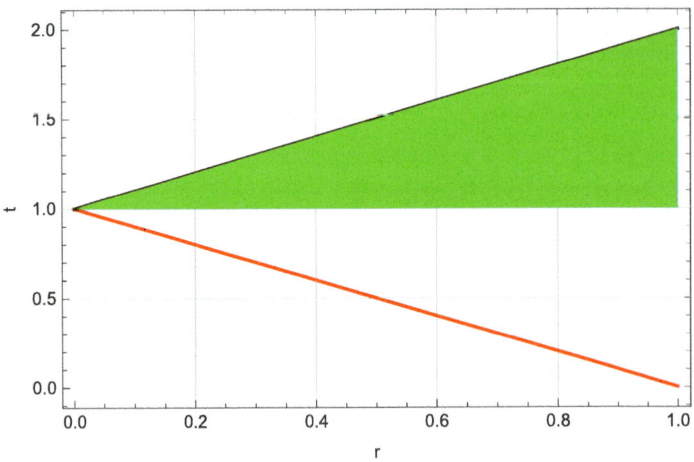

Figure A.2 The same as Fig. A.1, except for $s = 1$.

In the latter case, the integration region in the t–r plane is surrounded by four boundary lines $t = r + s, t = 1, r = s, r = 1$. It can be expressed as a trapezoid. As an example, the region for $s = 3/4$ is shown in green in Fig. A.5.

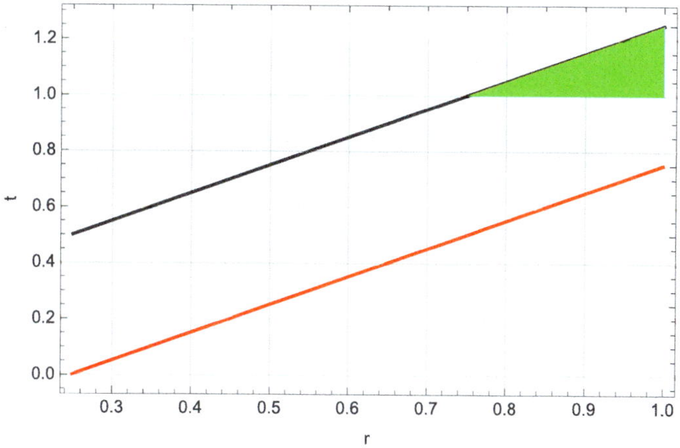

Figure A.3 Integration region (green triangle) in the t–r plane for $s = 1/4$ in the second term on the right side of Eq. (5.69). Black line: $t = r + s$, red line: $t = r - s$.

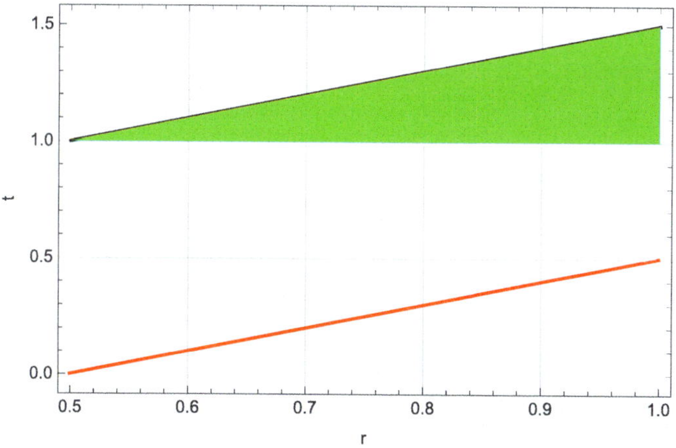

Figure A.4 The same as Fig. A.3, but for $s = 1/2$.

A.3 Integration Region 3

The integration region in the t–r plane for $0 < s < 1$ in the second term on the right-hand side of Eq. (5.68) can be represented by a pentagon surrounded by the five boundary lines $r = 1$, $r = R$

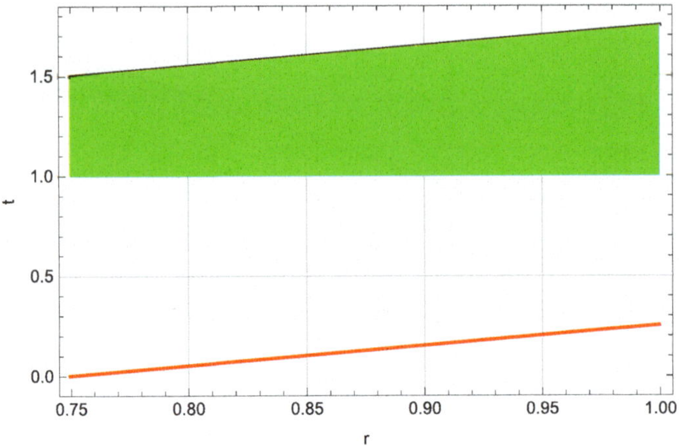

Figure A.5 Integration region (green trapezoid) in the t–r plane for $s = 3/4$ in the second term on the right-hand side of Eq. (5.69). Black line: $t = r + s$, red line: $t = r - s$.

$(R \to \infty)$, $t = r + s$, $t = r - s$, and $t = 1$. Here, as examples, those for $s = 1/2$ and $s = 1$ are shown in green in Fig. A.6 and Fig. A.7, respectively.

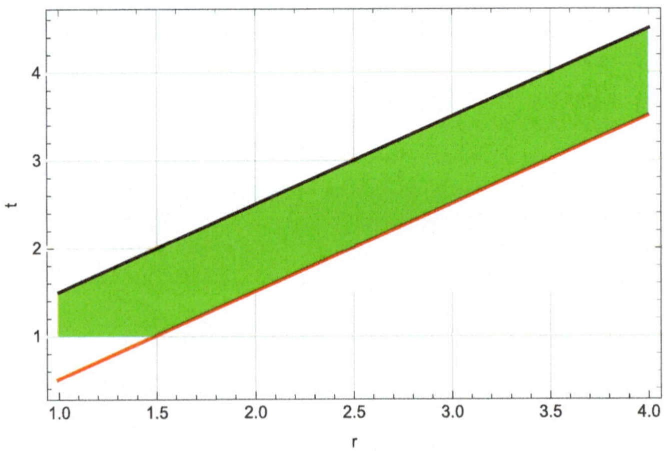

Figure A.6 Integration region (green pentagon) in the t–r plane for $s = 1/2$ in the second term on the right-hand side of Eq. (5.68). Black line: $t = r + s$, red line: $t = r - s$.

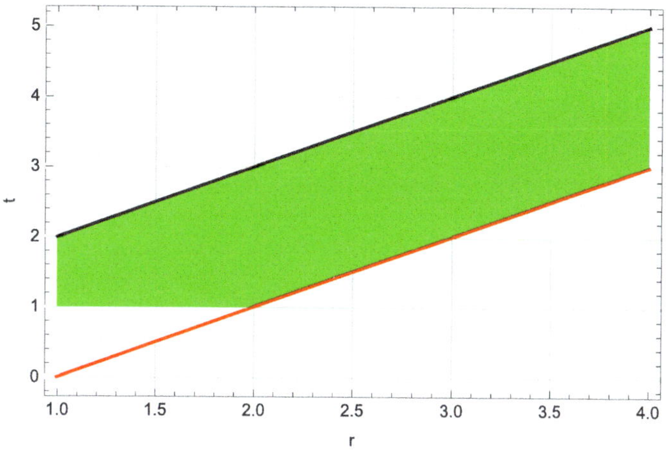

Figure A.7 The same as Fig. A.6, except for $s = 1$.

Appendix B

$\hat{Q}_0(n)$, $\mathscr{Q}_1(n, k)$, $\mathscr{Q}_2(n, k)$

The coefficients $\hat{Q}_0(n)$, $\mathscr{Q}_1(n, k)$, $\mathscr{Q}_2(n, k)$ $(1 \le n, k \le N)$ in Eqs. (5.117) are given by

$$\hat{Q}_0(k) = \frac{\pi}{z_k^3 \Delta}\left[2 - z_k^2 + \frac{\pi\rho}{2\Delta}(2 - z_k) - 2\exp(-z_k)\hat{a}(k)\right].$$

$$\mathscr{Q}_1(k, n) = a_n\left[\frac{1}{z_k z_n} + \hat{b}(k, n) - \hat{d}(k, n)\exp(-z_k)\right].$$

$$\mathscr{Q}_2(k, n) = \frac{a_n\rho}{z_n^2}\left[\hat{e}(k, n) - \hat{Q}_1(k, n)\right],$$

where,

$$\hat{Q}_1(k, n) = 2\exp(-z_n)\pi\left[\frac{1}{z_k z_n} + \hat{b}(k, n) - \hat{d}(k, n)\exp(-z_k)\right],$$

$$\hat{a}(k) = \frac{\pi\rho}{4\Delta}(z_k + 2) + z_k + 1,$$

$$\hat{b}(k, n) = \frac{\pi\rho}{z_k^3 z_n^3 \Delta}\left[\frac{\pi\rho}{2\Delta}(4 - 2z_k + 2z_n - z_k z_n) + 2z_k - 2z_k^2 - 2z_n + 2z_k z_n - z_k^2 z_n - 2z_n^2 + z_k z_n^2\right],$$

The Physics of Liquid Water
Makoto Yasutomi

ISBN 978-981-4877-25-1 (Hardcover), 978-1-003-05616-4 (eBook)
www.jennystanford.com

$$\hat{d}(k,n) = \frac{\pi\rho}{z_k^3 z_n^3 \Delta}\left[\frac{\pi\rho}{2\Delta}(4 + 2z_k + 2z_n + z_k z_n) + 2z_k - 2z_n - 2z_n^2 - z_k z_n^2\right] + \frac{1}{z_k(z_k+z_n)},$$

$$\hat{e}(k,n) = 2\pi\left\{ -\exp(-z_k)\,\hat{c}(k,n) + \frac{1}{z_n(z_k+z_n)} + \frac{\pi\rho}{z_k^3 z_n^3 \Delta}\left[\frac{\pi\rho}{2\Delta}(4 - 2z_k - 2z_n + z_k z_n) + 2z_k - 2z_k^2 - 2z_n + z_k^2 z_n\right]\right\},$$

$$\hat{c}(k,n) = \frac{\pi\rho}{z_k^3 z_n^3 \Delta}\left[\frac{\pi\rho}{2\Delta}(4 + 2z_k - 2z_n - z_k z_n) + 2z_k - 2z_n - 2z_k z_n\right].$$

Appendix C

$F_0(k)$, $F_1(k)$, $\mathscr{F}_2(k, n)$, $\mathscr{F}_3(k, n)$, $\mathscr{F}_4(k, n)$, $\mathscr{F}_5(k, n)$

The coefficients $F_0(k)$, $F_1(k)$, $\mathscr{F}_2(k, n)$, $\mathscr{F}_3(k, n)$, $\mathscr{F}_4(k, n)$, and $\mathscr{F}_5(k, n)$ in Eqs. (5.118) are given by the following equations:

$$F_0(k) = \frac{\hat{a}(k)}{z_k \Delta}, \quad F_1(k) = -1 + \rho \dot{Q}_0(k),$$

$$\mathscr{F}_2(k, n) = \frac{a_n}{2\pi} z_k^2 \hat{d}(k, n),$$

$$\mathscr{F}_3(k, n) = \frac{a_n \rho z_k^2}{z_n^2} \left[\hat{c}(k, n) - \hat{d}(k, n) \exp(-z_n) \right],$$

$$\mathscr{F}_4(k, n) = \rho \mathscr{Q}_1(k, n). \quad \mathscr{F}_5(k, n) = \rho \mathscr{Q}_2(k, n).$$

The Physics of Liquid Water
Makoto Yasutomi

ISBN 978-981-4877-25-1 (Hardcover), 978-1-003-05616-4 (eBook)
www.jennystanford.com

Appendix D

A, B

A and B included in $Q(r)$ are given by the following equations:

$$A = A_0 + \frac{\pi\rho}{\Delta}\sum_{n=1}^{N}\frac{a_n}{z_n^3}\left[A_1(n) + A_2(n)\mathscr{G}_n\right]\mathscr{D}_n. \tag{D.1}$$

$$B = B_0 + \frac{\pi\rho}{2\Delta}\sum_{n=1}^{N}\frac{a_n}{z_n}\left[B_1(n) + B_2(n)\mathscr{G}_n\right]\mathscr{D}_n. \tag{D.2}$$

Where, A_0, $A_1(n)$, $A_2(n)$, B_0, $B_1(n)$, $B_2(n)$ are given by the following equations:

$$A_0 = \frac{2\pi}{\Delta}\left(1 + \frac{\pi\rho}{2\Delta}\right), \quad A_1(n) = \frac{\pi\rho}{\Delta}(2 + z_n) - 2z_n(1 + z_n),$$

$$A_2(n) = -\frac{2\pi\rho}{z_n^2}\left[A_1(n)\exp(-z_n) + \frac{\pi\rho}{\Delta}(z_n - 2) + 2z_n\right],$$

$$B_0 = -\frac{\pi^2\rho}{2\Delta^2}, \quad B_1(n) = \frac{2\left(z_n^2 + 2z_n + 2\right)}{z_n^2} - \frac{\pi\rho}{\Delta}\frac{z_n + 2}{z_n^2},$$

$$B_2(n) = \frac{2\pi\rho}{z_n^2}\left\{-\exp(-z_n)B_1(n) + \frac{1}{z_n^2}\left[4 + \frac{\pi\rho}{\Delta}(z_n - 2)\right]\right\}.$$

The Physics of Liquid Water
Makoto Yasutomi

ISBN 978-981-4877-25-1 (Hardcover), 978-1-003-05616-4 (eBook)
www.jennystanford.com

Appendix E

$\partial x_j/\partial u$

In equations $\mathscr{F}_k = 0$ $(1 \le k \le 2N)$, let dx_j be the slight changes of x_j $(1 \le j \le 2N)$ which are associated with the small change du in u, we get following equations:

$$d\mathscr{F}_k = \frac{\partial \mathscr{F}_k}{\partial u} du + \sum_{j=1}^{2N} \frac{\partial \mathscr{F}_k}{\partial x_j} dx_j = 0. \quad \therefore \quad \sum_{j=1}^{2N} \frac{\partial \mathscr{F}_k}{\partial x_j} \frac{\partial x_j}{\partial u} = -\frac{\partial \mathscr{F}_k}{\partial u}.$$

Therefore, $\partial x_j/\partial u$ $(1 \le j \le 2N)$ are obtained by solving the following $2N$ simultaneous linear equations:

$$\begin{bmatrix} 0 & \frac{\partial \mathscr{F}_1}{\partial x_2} & \cdots\cdots & \frac{\partial \mathscr{F}_1}{\partial x_N} & 0 & \cdots & 0 \\ \frac{\partial \mathscr{F}_2}{\partial x_1} & \frac{\partial \mathscr{F}_2}{\partial x_2} & \cdots\cdots & \cdots & \cdots\cdots & & \frac{\partial \mathscr{F}_2}{\partial x_{2N}} \\ \frac{\partial \mathscr{F}_3}{\partial x_1} & \frac{\partial \mathscr{F}_3}{\partial x_2} & \cdots\cdots & \cdots & \cdots\cdots & & \frac{\partial \mathscr{F}_3}{\partial x_{2N}} \\ \vdots & & & & & & \\ \frac{\partial \mathscr{F}_{2N}}{\partial x_1} & \frac{\partial \mathscr{F}_{2N}}{\partial x_2} & \cdots\cdots & \cdots & \cdots\cdots & & \frac{\partial \mathscr{F}_{2N}}{\partial x_{2N}} \end{bmatrix} \begin{bmatrix} \frac{\partial x_1}{\partial u} \\ \frac{\partial x_2}{\partial u} \\ \frac{\partial x_3}{\partial u} \\ \vdots \\ \frac{\partial x_{2N}}{\partial u} \end{bmatrix} = \begin{bmatrix} -1 \\ 0 \\ 0 \\ \vdots \\ 0 \end{bmatrix}.$$

The Physics of Liquid Water
Makoto Yasutomi

ISBN 978-981-4877-25-1 (Hardcover), 978-1-003-05616-4 (eBook)
www.jennystanford.com

Appendix F

Stability of Solutions of Diffusion Equations

It is explained that diffusion type equations or heat conduction type equations (5.124) have stable solutions when $\mathcal{B}(\rho, u)$ is positive, but are unstable when negative. Here, for simplicity, the case where $\rho/\mathcal{B}(\rho, u) = k$ is a constant is described. That is,

$$\frac{\partial u(\rho, \beta)}{\partial \beta} = k \frac{\partial^2 u(\rho, \beta)}{\partial \rho^2}. \qquad \text{(F.1)}$$

As boundary conditions, the following equations are considered:

$$u(0, \beta) = \frac{\partial^2 u(1, \beta)}{\partial \rho^2} = 0 \quad (\beta \geq 0). \qquad \text{(F.2)}$$

As initial conditions, consider the following equations:

$$u(\rho, 0) = f(\rho), \; f(0) = 0, \; f''(1) = 0, \qquad (0 \leq \rho \leq 1). \qquad \text{(F.3)}$$

We put

$$u(\rho, \beta) = R(\rho)T(\beta), \qquad \text{(F.4)}$$

where, $R(\rho)$ is an unknown function of ρ and $T(\beta)$ is an unknown function of β. Substitute Eq. (F.4) into Eq. (F.1), and divide both sides of the equation by kRT to get

$$\frac{1}{kT}\frac{dT}{d\beta} = \frac{1}{R}\frac{d^2R}{d\rho^2} = \lambda.$$

The Physics of Liquid Water
Makoto Yasutomi

ISBN 978-981-4877-25-1 (Hardcover), 978-1-003-05616-4 (eBook)
www.jennystanford.com

Here, since the left side is a function of only β and the right side is a function of only ρ, it turns out that λ must be constant. Therefore, the above equation can be separated into the following two equations:

$$\frac{dT}{d\beta} = k\lambda T, \qquad \frac{d^2R}{d\rho^2} = \lambda R \qquad (\lambda = \text{const}). \tag{F.5}$$

First, solve the second equation of (F.5). If we set $R = e^{s\rho}$, we get

$$(s^2 - \lambda)e^{s\rho} = 0 \quad \rightarrow \quad s^2 = \lambda. \tag{F.6}$$

F.1 In the Case of $\lambda > 0$

From Eq. (F.6), we get $s = \pm\sqrt{\lambda}$, and two independent solutions $e^{\sqrt{\lambda}\rho}$ and $e^{-\sqrt{\lambda}\rho}$. Therefore, the general solution is

$$R = A_1 e^{-\sqrt{\lambda}\rho} + A_2 e^{\sqrt{\lambda}\rho} \qquad (A_1, A_2 : \text{any constants}).$$

From the boundary conditions (F.2)

$$A_1 + A_2 = 0, \quad \lambda(A_1 e^{-\sqrt{\lambda}} + A_2 e^{\sqrt{\lambda}}) = 0.$$

Solving the above equation yields $A_1 = A_2 = 0$, so for any (ρ, β), $u(\rho, \beta) = 0$. It is unsuitable as a solution.

F.2 In the Case of $\lambda = 0$

From the second expression of Eqs. (F.5)

$$\frac{d^2R}{d\rho^2} = 0 \quad \rightarrow \quad R = B_1\rho + B_2.$$

Substituting this into the boundary condition (F.2)

$$B_2 = 0, \quad B_1 = \text{constant}.$$

Also, because $T = $ constant is obtained from the first expression of (F.5). we get

$$u(\rho, \beta) = RT = D\rho \quad (D : \text{constant}).$$

F.3 In the Case of λ < 0

From (F.6), two independent solutions $e^{-i\sqrt{-\lambda}\rho}$ and $e^{i\sqrt{-\lambda}\rho}$ are obtained. In this case, $\cos\sqrt{-\lambda}\rho = (e^{i\sqrt{-\lambda}\rho} + e^{-i\sqrt{-\lambda}\rho})/2$ and $\sin\sqrt{-\lambda}\rho = (e^{i\sqrt{-\lambda}\rho} - e^{-i\sqrt{-\lambda}\rho})/(2i)$ are another two independent solutions, so the general solution is

$$R = C_1 \cos\sqrt{-\lambda}\rho + C_2 \sin\sqrt{-\lambda}\rho \qquad (C_1, C_2 : \text{constants}).$$

From the above equation and the boundary conditions (F.2),

$$C_1 = 0, \quad C_2\lambda \sin\sqrt{-\lambda} = 0.$$

If $C_2 = 0$, for any (ρ, β), $u(\rho, \beta) = 0$, which is not suitable as a solution. Therefore,

$$\sqrt{-\lambda} = n\pi \quad (n = 1, 2, 3, \cdots).$$

From the first expression of (F.5)

$$\frac{dT}{T} = k\lambda d\beta \quad \rightarrow \quad \ln T = k\lambda\beta + E \quad \rightarrow \quad T = e^E e^{-k(n\pi)^2\beta}$$

where, E is constant. In summary,

$$u(\rho, \beta) = RT = C' e^{-k(n\pi)^2\beta} \sin n\pi\rho \quad (C' = e^E C_2 : \text{constant}).$$

F.4 Principle of Superposition of Solutions of Diffusion Equation

For any integer N, if u_n $(n = 1, 2, 3, \cdots, N)$ are the solutions of the diffusion equation (F.1), the combined $u = \sum_{n=1}^{N} c_n u_n$ (c_n: constants) is also a solution (superposition principle). See the proof as follows:

Proof. From the above assumption,

$$\frac{\partial u_n}{\partial \beta} = k\frac{\partial^2 u_n}{\partial \rho^2} \quad (n = 1, 2, 3, \cdots, N).$$

So substituting $u = \sum_{n=1}^{N} c_n u_n$ into Eq. (F.1), we get

$$\frac{\partial u}{\partial \beta} = \sum_{n=1}^{N} c_n \frac{\partial u_n}{\partial \beta} = k\frac{\partial^2}{\partial \rho^2}\sum_{n=1}^{N} c_n u_n = k\frac{\partial^2 u}{\partial \rho^2}.$$

Therefore, u is the solution of Eq. (F.1).

To summarize the results of Sections F.1, F.2, F.3, and F.4, the general solution of Eq. (F.1) is

$$u(\rho, \beta) = c_0\rho + \sum_{n=1}^{N} c_n e^{-k(n\pi)^2\beta} \sin n\pi\rho. \tag{F.7}$$

Here, c_0 and c_n $(n = 1, 2, 3, \cdots, N)$ are arbitrary constants. These constants can be uniquely determined as follows by specifying $f(\rho)$ of the initial condition (F.3). First, from $u(\rho, 0) = f(\rho)$,

$$c_0\rho + \sum_{n=1}^{N} c_n \sin n\pi\rho = f(\rho). \tag{F.8}$$

If both sides of Eq. (F.8) are integrated over ρ in the interval of $0 \leq \rho \leq 1$,

$$\frac{1}{2}c_0 + \frac{2}{\pi}\sum_{m=1}^{m\leq(N+1)/2} \frac{c_{2m-1}}{2m-1} = \int_0^1 f(\rho)d\rho. \tag{F.9}$$

Multiplying both sides of (F.8) by $\sin m\pi\rho$ and integrating with respect to ρ in the range $0 \leq \rho \leq 1$,

$$\frac{c_0}{m\pi}(-1)^{m+1} + \frac{c_m}{2} = \int_0^1 f(\rho)\sin m\pi\rho\, d\rho, \qquad (m = 1, 2, 3, \cdots, N). \tag{F.10}$$

If you solve Eqs. (F.9) and (F.10) simultaneously, c_n $(n = 0, 1, 2, \cdots, N)$ can be obtained.

As can be seen from the general solution (F.7), if $k < 0$, the solution $u(\rho, \beta)$ becomes infinite as β increases. You can see that it diverges.

Appendix G

Solutions of $\mathscr{D}_n$ and $\mathscr{G}_n$ ($1 \leq n \leq 6$) for Water below T_c for ϕ_{c1}

G.1 Solutions of $\mathscr{D}_n$ and $\mathscr{G}_n$ ($1 \leq n \leq 6$) for Gas-Phase Water

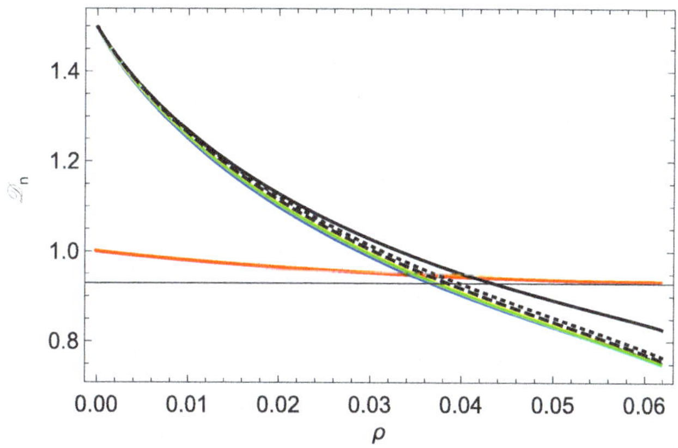

Figure G.1 $\mathscr{D}_n(\rho, \beta)$ $(1 \leq n \leq 6)$ of gas phase water at $\beta = 0.1154$. Red line: $\mathscr{D}_1$, Black line: $\mathscr{D}_2$, Blue line: $\mathscr{D}_3$, Green line: $\mathscr{D}_4$, Dashed line: $\mathscr{D}_5$, Dotted line: $\mathscr{D}_6$.

The Physics of Liquid Water
Makoto Yasutomi

ISBN 978-981-4877-25-1 (Hardcover), 978-1-003-05616-4 (eBook)
www.jennystanford.com

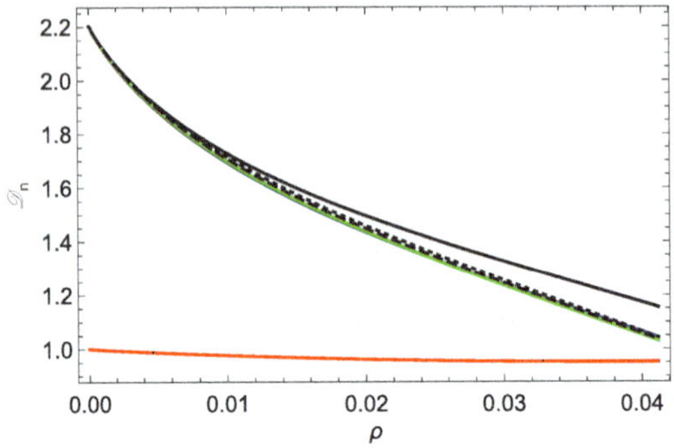

Figure G.2 The same as Fig. G.1, but for $\beta = 0.1414$.

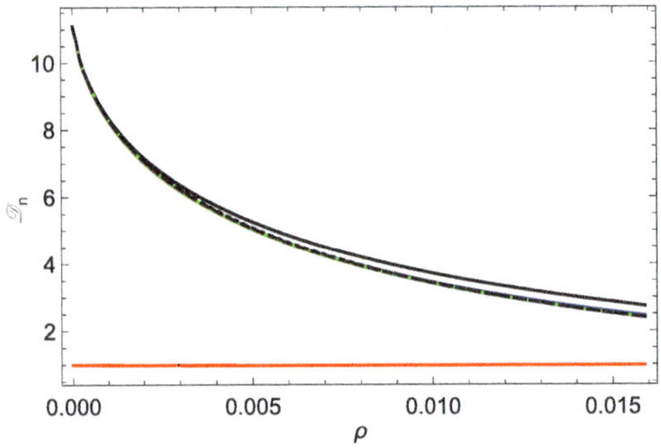

Figure G.3 The same as Fig. G.1, but for $\beta = 0.2684$.

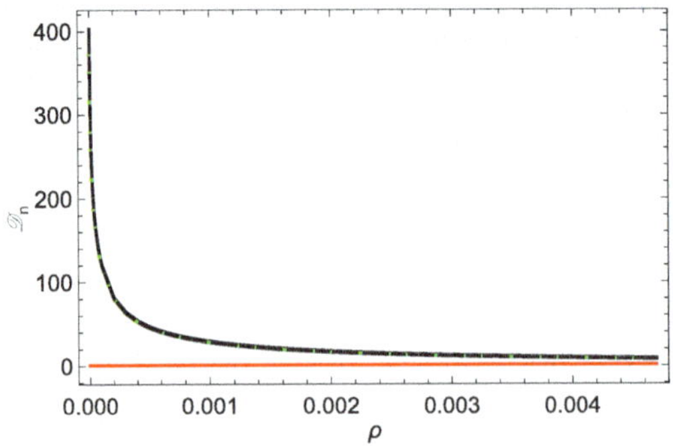

Figure G.4 The same as Fig. G.1, but for $\beta = 0.5954$.

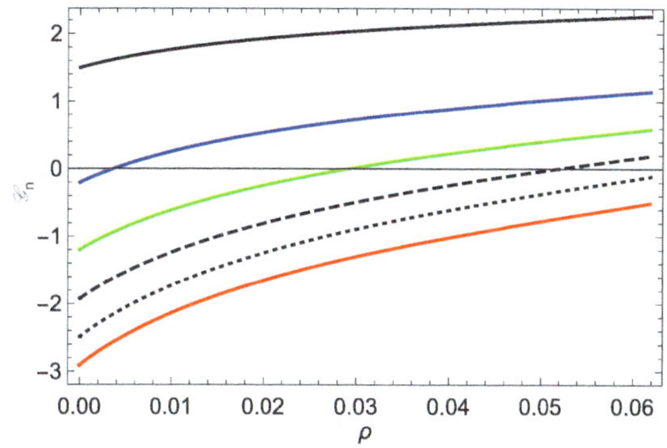

Figure G.5 The same as Fig. G.1, but for $\mathscr{G}_n$ ($1 \leq n \leq 6$).

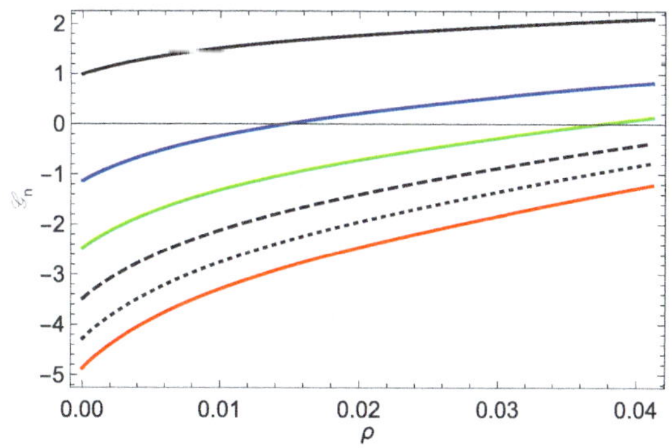

Figure G.6 The same as Fig. G.2, but for $\mathscr{G}_n$ ($1 \leq n \leq 6$).

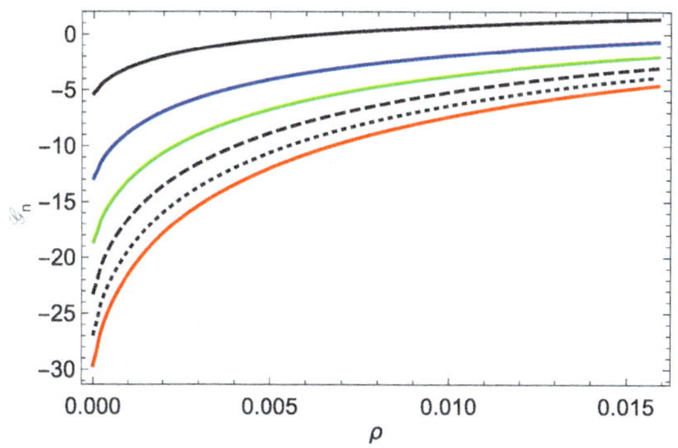

Figure G.7 The same as Fig. G.3, but for $\mathscr{G}_n$ ($1 \leq n \leq 6$).

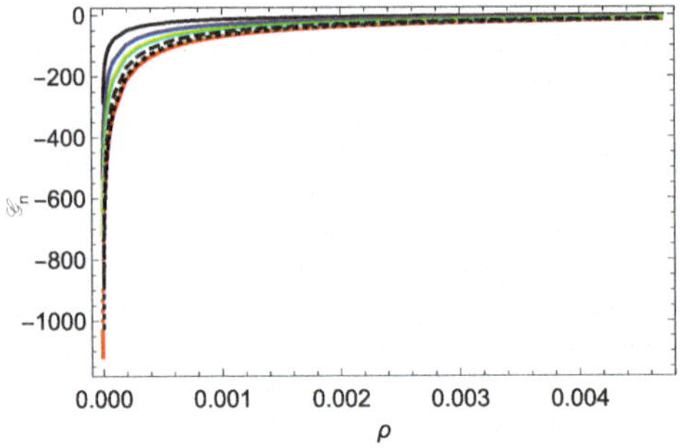

Figure G.8 The same as Fig. G.4, but for $\mathscr{G}_n$ $(1 \leq n \leq 6)$.

G.2 Solutions of $\mathscr{D}_n$ and $\mathscr{G}_n$ $(1 \leq n \leq 6)$ for Liquid-Phase Water

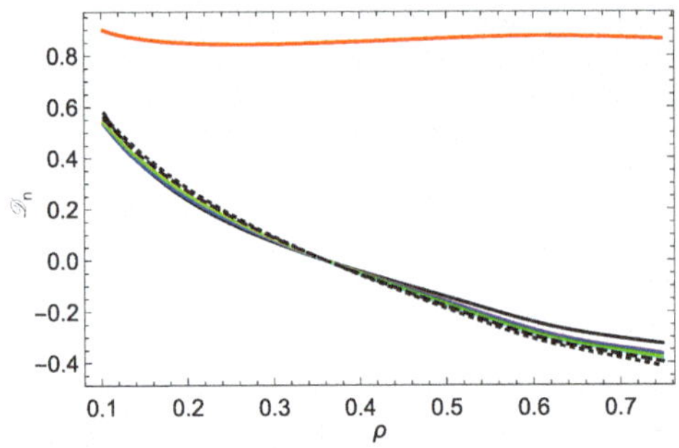

Figure G.9 The same as Fig. G.1, but for liquid water.

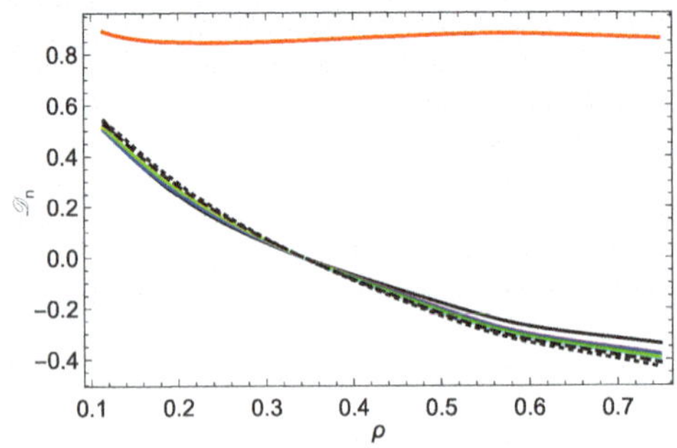

Figure G.10 The same as Fig. G.2, but for liquid water.

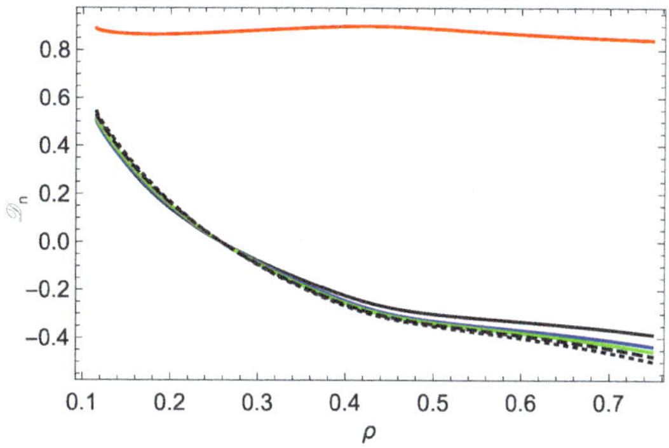

Figure G.11 The same as Fig. G.3, but for liquid water.

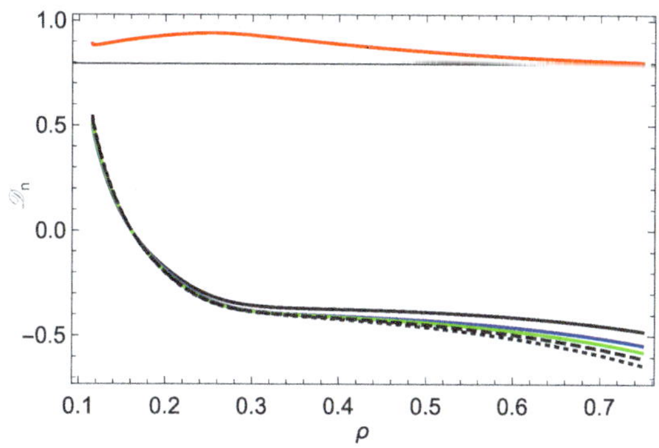

Figure G.12 The same as Fig. G.4, but for liquid water.

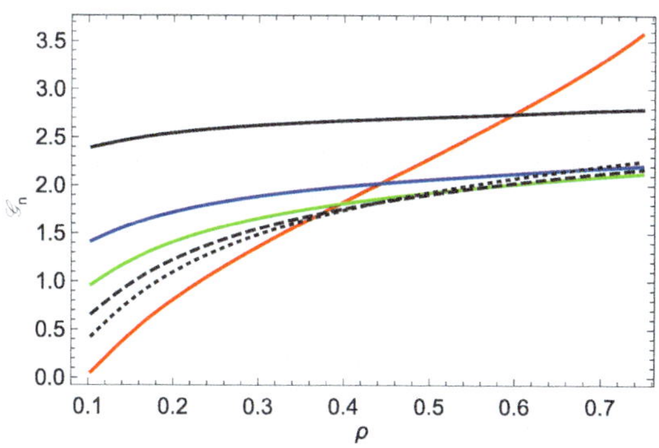

Figure G.13 The same as Fig. G.5, but for liquid water.

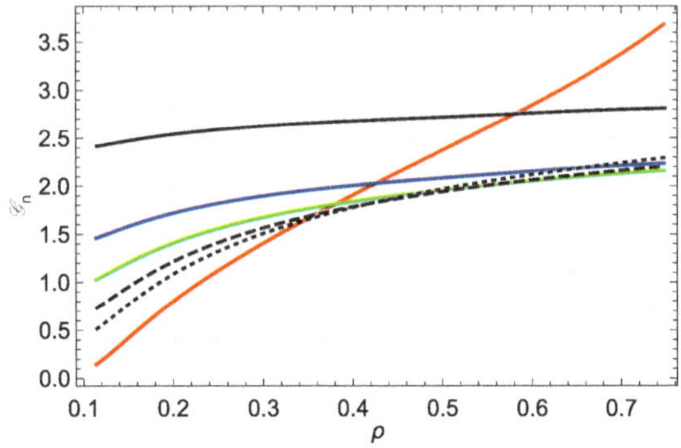

Figure G.14 The same as Fig. G.6, but for liquid water.

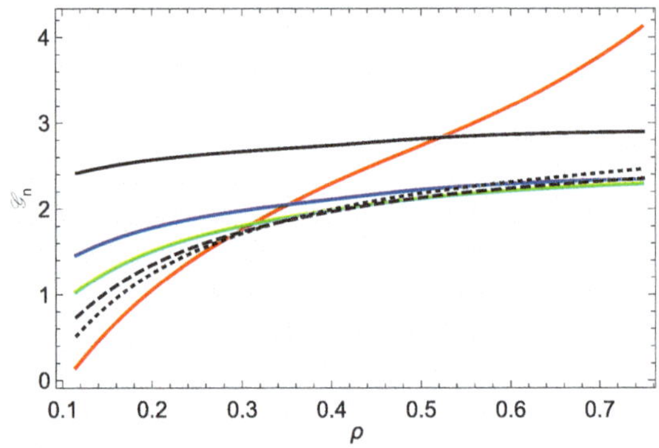

Figure G.15 The same as Fig. G.7, but for liquid water.

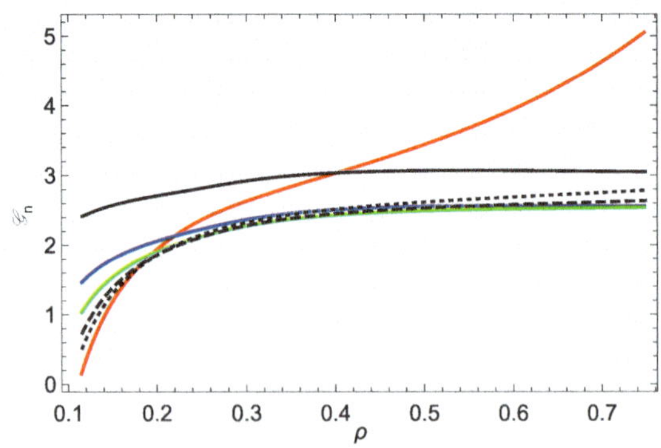

Figure G.16 The same as Fig. G.8, but for liquid water.

Bibliography

Abascal, J. L. F., and Vega, C. A general purpose model for the condensed phases of water: TIP4P/2005, *J. Chem. Phys.*, **123**, 234505 (2005).

Barboy, B., and Tenne, R. *Chem. Phys.*, **38**, 369 (1979).

Baxter, R. J. *Australian J. Phys.*, **21**, 563 (1968).

Baxter, R. J. *J. Chem. Phys.*, **49**, 2770 (1968).

Baxter, R. J. *J. Chem. Phys.*, **52**, 4559 (1970).

Benedict, W. S., Gailar, N., and Plyler, E. K. *J. Chem. Phys.*, **24**, 1139 (1956).

Berendsen, H. J. C., Grigera, J. R., and Straatsma. T. P. The missing term in effective pair potentials, *J. Phys. Chem.*, **91**, 6269 (1987).

Bernal, J. D., and Fowler, R. H. A theory of water and ionic solution, with particular reference to hydrogen and hydroxyl ions, *J Chem Phys*, **1**, 515–548 (1933). doi:10.1063/1.1749327.

Betancourt-Cárdenas, F. F. Galicia-Luna, L. A., Benavides, A. L., Ramirez, J. A., and Schöll-Paschinger, E. Thermodynamics of a long-range triangle-well fluid, *Mol. Phys.*, **106**, 113 (2008).

Blum, L. *J. Statist. Phys.*, **22**, 661 (1980).

Blum, L., and Høye, J. S. *J. Statist. Phys.*, **19**, 317 (1978).

Caballero, J. B., and Puertas, A. M. Density anomaly and liquid–liquid transition from perturbation theories, *Phys. Rev. E*, **74**, 051506 (2006).

Carnahan, N. F., and Starling, K. E. *J. Chem. Phys.*, **51**, 635 (1969). For a listing of other hard-sphere equations of state and an assessment of their relative merits, see Murelo, A., Galán, C. A., Parra, M. I., and Cuadros, F., Lect. *Notes Phys*, **753**, 37 (2008).

del Río, F., Ávalos, E., Espíndola, R., Rull, L. F., Jackson, G., and Lago, S. *Mol. Phys.*, **100**, 2531 (2002).

Duda, Y. *J. Chem. Phys.*, **130**, 116101 (2009).

Eisenberg, D., and Kauzmann, W. J. *The Structure and Properties of Water*, London: Oxford University Press, Inc. (1969).

Elliott, J. R., and Hu, L. *J. Chem. Phys.*, **110**, 3043 (1999).

Fomin, Y. D., Tsiok, E. N., and Ryzhov, V. N. Inversion of sequence of diffusion and density anomalies in core-softened systems, *J. Chem. Phys.*, **135**, 234502 (2011).

Gibson, H. M., and Wilding, N. B. Metastable liquid–liquid coexistence and density anomalies in a core-softened fluid, *Phys. Rev. E*, **73**, 061507 (2006)

Ginoza, M. *J. Phys. Soc. Japan*, **54**, 2783 (1985).

Ginoza, M. *J. Phys. Soc. Japan*, **55**, 95 (1986).

Ginoza, M., and Yasutomi, M. *Mol. Phys.*, **87**, 593 (1996).

Gribova, N. V., Fomin, Y. D., Frenkel, D., and Ryzhov, V. N. Water like thermodynamic anomalies in a repulsive-shoulder potential system, *Phys. Rev. E*, **79**, 051202 (2009).

Hansen, P., and McDonald, I. R. *Theory of Simple Liquids*, 2nd edition, New York: Academic Press (1986).

Henderson, D., Abraham, F. F., and Barker, J. A. *Mol. Phys.*, **31**, 1291 (1976).

Henderson, D., Lebowitz, J. L., Blum, L., and Waisman, E. *Mol. Phys.*, **39**, 47 (1980).

Høy, J. S., and Blum, L. *J. Statist. Phys.*, **16**, 399 (1977).

Jagla, E. A. Low-temperature behavior of core-softened models: water and silica behavior, *Phys. Rev. E*, **63**, 061509 (2001).

Jorgensen, W. L., Chandrasekhar, J., Madura, J. D., Impey, R. W., and Klein, M. L. Comparison of simple potential functions for simulating liquid water, *J. Chem. Phys.*, **79**, 926 (1983).

Lide, D. R. *CRC Handbook of Chemistry and Physics*, 70th edition, Boca Raton (FL): CRC Press (1990).

Liu, H., Garde, S., and Kumar, S. *J. Chem. Phys.*, **123**, 174505 (2005).

Lomba, E., Almarza, N. G., Martin, C., and McBridge, C. Phase behavior of attractive and repulsive ramp fluids: integral equation and computer simulation studies, *J. Chem. Phys.*, **126**, 244510 (2007).

López-Rendóu, R., Reyes, Y., and Orea, P. *J. Chem. Phys.*, **125**, 084508 (2006).

Mahoney, M. W., and Jorgensen, W. L. A five-site model for liquid water and the reproduction of the density anomaly by rigid, nonpolarizable potential functions, *J. Chem. Phys.*, **112**, 8910 (2000).

Malescio, G., Prestipino, S., and Saija, F. Anomalous melting and solid polymorphism of a modified inverse-power potential, *Mol. Phys.*, **109**, 2837–2844 (2011).

Mallamace, F., Branca, C., Broccio, M., Corsaro, C., Mou, C. Y., and Chen, C. Y. *Proc Natl Acad Sci USA*, **104**, 18387–18391 (2007).

Mayer, M., and Stanley, H. E. Liquid–liquid phase transition in confined water: A Monte Carlo study, *J. Phys. Chem. B*, **103**, 9728–9730 (1999). doi: 10.1021/jp984142f.

Mishima, O., and Stanley, H. E. Decompression-induced melting of ice IV and the liquid–liquid transition in water, *Nature*, **392**, 164–168 (1998).

Molinero, V., and Moore, E. B. Water modeled as an intermediate element between carbon and silicon, *J. Phys. Chem. B*, **113**, 4008 (2009).

Nobel, B. *Methods on the Wiener-Hopf Technique*, New York: Pergamon Press, Inc. (1958), p. 157.

Nowak, E. S., Grosh, R. J., and Liley, P. E. *J. Heat Transfer*, **83C**, 1 (1961).

Oliveira, A. B. De, Netz, P. A., and Barbosa, M. C. An ubiquitous mechanism for water-like anomalies, *EPL*, **85**, 36001 (2009)

Pagan, D. L., and Gunton, J. D. *J. Chem. Phys.*, **122**, 184515 (2005).

Paley, R. E. A. C., and Wiener, N. *Fourier Transforms in the Complex Domain*, Colloquim Publications/American Mathematical Society, V. 19 (1934).

Pauling, L. *J. Am. Chem. Soc.*, **57**, 2680 (1935).

Pauling, L. The structure of water, In *Hydrogen Bonding*, D. Hadzi and H. W. Thompson (eds.), London: Pergamon Press Ltd (1959), p. 1.

Perram, J. W., and Smith, E. R. *Chem. Phys. Lett.*, **35**, 138 (1975).

Pini, D., Stell, G., and Wilding, N. B. A liquid-state theory that remains successful in the critical region, *Mol. Phys.*, **95**, 483 (1998).

Pini, D., Stell, G., and Dickman, R. Thermodynamically self-consistent theory of structure for three-dimensional lattice gases, *Phys. Rev. E*, **57**, 2862 (1998).

Pini, D., Stell, G., and Wilding, N. B. *Molecular Physics*, **95**, 483–494 (1998).

Plischke, M., and Henderson, D. *Proc. R. Soc. A*, **404**, 323 (1986).

Poole P. H., Sciortino, R., Essmann, U., and Stanley, H. E. Phase behaviour of metastable water, *Nature*, **360**, 324–328 (1992). doi: 10.1038/360324a0.

Pople, J. A. Molecular association in liquids. II. a theory of the structure of water, *Proc. R. Soc. Lond. A*, **205**, 163 (1951).

Prestipino, S., Saija, F., and Malescio, G. Anomalous phase behavior in a model fluid with only one type of local structure, *J. Chem. Phys.*, **133**, 144504 (2010).

Röntgen W. C. Ueber die Constitution des flüssigen Wassers. *Annalen der Physik*. (1892) **281**:91-7. doi:10.1002/andp.18922810108.

Russo, J., and Tanaka, H. Understanding water's anomalies with locally favoured structures, *Nat Commun.*, **5**, 3556 (2014).

Sadr-Lahijany, M. R., Scala, A., Buldyrev, S. V., and Stanley, H. E. Liquid-state anomalies and the Stell-Hemmer core-softened potential, *Phys. Rev. Lett.*, **81**, 4895 (1998).

Samoilov, O. Y. A new approach to the study of hydrogen of ions in aqueous solutions, *Discuss. Faraday Soc.*, **24**:141–146 (1957). doi:10.1039/df9572400141.

Schöll-Paschinger, E. Self-consistent Ornstein–Zernike approximation for the Sogami-Ise fluid, *J. Chem. Phys.*, **120**, 11698 (2004).

Schöll-Paschinger, E. Vapor–liquid equilibrium and critical behavior of the square-well fluid of variable range: A theoretical study, *J. Chem. Phys.*, **123**, 234513 (2005).

Sciortino, F., Nave E. L., and Tartaglia, P. Physics of the liquid-liquid critical point, *Phys. Rev. Lett.*, **91**, 155701-1 (2003). doi: 10.1103/PhysRevLett.91.155701.

Skibinsky, A., Buldyerv, S. V., Franzese, G., Malescio, G., and Stanley, H. E. *Phys. Rev. E*, **69**, 061206 (2004).

Sun, Z. R., Sun. G., Chen, Y. X., et al. Liquid-liquid phase transition in water, *Sci China-Phys Mech Astron*, **57**, 810–818 (2014). doi: 10.1007/s11433-014-5451-z.

Thiele, E. *J. Chem. Phys*, **39**, 474 (1963).

Thompson, N. E., Isbister, D. J., Bearman, R. J., and Freasier, B. C. *Mol. Phys.*, **39** 27 (1980).

Waisman, E. *Mol. Phys.*, **25**, 4 (1973).

Waisman, E., Henderson, D., and Lebowitz, J. L. *Mol. Phys.*, **32**, 1373 (1976).

Wertheim, M. S. *Phys. Rev. Lett.*, **10**, 321 (1963).

Wertheim, M. S. *J. Math. Phys.*, **5**, 643 (1964).

Wilding, N. B., and Magee, J. E. Phase behavior and thermodynamic anomalies of core-softened fluids, *Phys. Rev. E*, **66**, 031509 (2002).

Yan, Z., Buldyrev, S. V., Giovambattista, N., and Stanley, H. E. Structural order for one-scale and two-scale potentials, *PRL*, **93**, 130604 (2004).

Yan, Z., Buldyrev, S. V., Kumar, P., Giovambattista, N., and Stanley, H. E. Correspondence between phase diagrams of the TIP5P water model and a spherically symmetric repulsive ramp potential with two characteristic length scales, *Phys. Rev. E*, **77**, 042201 (2008).

Yasutomi, M. A self-consistent Ornstein-Zernike approximation for a fluid with a screened power series interaction, *J. Chem. Phys.*, **133**, 154115 (2010).

Yasutomi, M. Interparticle interactions between water molecules, *Front. Phys.*, **2**, 64 (2014).

Yasutomi, M. Thermodynamic mechanism of density anomaly of liquid water, *Front. Phys.*, **3**, 8 (2015).

Yasutomi, M., and Ginoza, M. *Mol. Phys.*, **89**, 1755 (1996).

Yasutomi, M., and Ginoza, M. *J. Phys.: Condens. Matter*, **12**, L605 (2000).

Yasutomi, M. *J. Phys.: Condens. Matter*, **13**, L255 (2001).

Yasutomi, M. Which shape characteristics of the intermolecular interaction of liquid water determine its compressibility, *Front. Phys.*, **4**, 21 (2016).

Vega, C., Abascal, J. L. F., Conde, M. M., and Aragones, J. L. *Faraday Discuss.*, **141**, 251–276 (2009).

Vega, L., de Miguel, E., Rull, L. F., Jackson, G., and McLure, I. A. *J. Chem. Phys.*, **96**, 2296 (1992).

Index